Growing ORCHIDS

The HYBRID Story

By the same author

GROWING ORCHIDS:

Book One, Cymbidiums and Slippers
Book Two, The Cattleyas and Other Epiphytes
Book Three, Vandas, Dendrobiums and Others
Book Four, The Australasian Families
The Specialist Orchid Grower
Expanding Your Orchid Collection

Growing ORCHIDS

The HYBRID Story

J. N. RENTOUL

LOTHIAN BOOKS

James N. Rentoul
25 December 1909 – 15 March 1991

Jim Rentoul died in early 1991 when this, the final volume of his Growing Orchids series, was near completion. It is the last of seven titles begun in 1980 with Growing Orchids Book One, *Cymbidiums and Slippers*. Although his personal store of knowledge has now been lost, it is fortunate that he chose to share his lifetime's experience of orchid growing with others through his writing. We, his children, have finished this volume to complete his task.

Alex Rentoul, Barbara Skilbeck, Joan O'Sullivan and Noelle Weatherley

A Lothian Book
Lothian Publishing Company Pty Ltd
11 Munro Street, Port Melbourne, Victoria 3207

First published 1991

National Library of Australia
Cataloguing in publication data:

Rentoul, J. N. (James Noel), 1909–1991
Growing orchids.

ISBN 0 85091 442 6.

1. Orchid culture. 2. Orchids.
I. Title.

635.93415

Typeset in Palatino by Typeset Gallery Sdn Bhd
Produced by Island Graphics
Printed in Singapore

FRONT COVER PICTURE: *Cattleya* Golden Talisman

Contents

Introduction

In the books in the Growing Orchids series and in other publications by the same author, the name James Dominy is mentioned frequently. In the nineteenth century he began hybridising a flora represented in almost every land mass — the orchids. When he did so he must have had little idea that what he had started would become a preoccupation of so many people.

One might imagine that the possibilities of these plants for further innovation would be exhausted after almost 150 years of work by so many of 'nature's imitators'. Perhaps 'imitators' is too harsh a term, however, to apply to those who take the trouble to master cross-pollinating techniques, understand what to do with the seeds and finally to grow their additions to a catalogue filled to overflowing.

Most growers would be incredulous if told that the number of hybrids rejected, discarded or otherwise disposed of from cross-pollinations in the last 100-odd years would be about 50 per cent. That is perhaps a moderate estimate too, if we take even a cursory look at the seven volumes of *Sander's List of Orchid Hybrids* which record all, or almost all, cross-pollinations registered from James Dominy's first production onwards.

This list burgeoned into a confusing number of alliances between genera and the exhilaration of hybridists in those late years of the nineteenth century still infects tyro and fanatic alike in the closing years of the twentieth.

Considering the extent to which the hybrid list has lengthened in the past two decades one is led to believe that few combinations remain unexplored, but one must still ask the question: Is there anything left to accomplish? Hybrids embracing six genera, though uncommon, are listed in the catalogue and the shadow of genetic interference is hanging over all the plants, threatening to accomplish what cross-pollination could not. This becomes more real each day.

In the four years following publication of the 1981–1985 list fifty-two new combinations of genera were registered, fairly evenly spread over that period and including innovations which could have only botanical interest and no lasting significance for the life-blood of orchid growing.

What is this life-blood? Is it that growers continually buy fresh stock? Why do we discard or kill off so many of the plants we buy?

One could think of many answers to those questions, but the principal answers may be found in two aspects of human behaviour. The first is desire, the second is the inability to find a satisfactory means of cultivating plants which we should never have bought in the first place without considering the available environment. So many of us persist to the stage where recovery of debilitated plants becomes impossible — they should have been handed over to other growers better situated before reaching that stage.

Since it is impossible to thoroughly review in one book a subject so extensive as the catalogue of orchid hybrid population, much has necessarily been omitted.

Nevertheless, an attempt has been made to convey an understanding of the past. Without that there is really no future. And the future — the twenty-first century — what does it offer that has not been found? Assessing that proposition by growers of the end of the twentieth century would be no more difficult than for growers at the beginning of that period. And how many of them could have guessed that orchids would be multiplied by the means we have found?

Botany and taxonomy are sometimes at odds with *Sander's List of Orchid Hybrids* and this is explained through this book. Sometimes the hybrid list and literature are at odds and this should be noted.

The work of registration authorities cannot be too highly praised, since they must oversee each entry, catalogue it, research it and finally approve it for inclusion in the list published monthly in the *Orchid Review*, the *American Orchid Review* and who knows how many other publications. At times, changes are made to names of genera and species. But once having appeared as a registered name in *Sander's List of Orchid Hybrids* that name prevails and taxonomic or botanic alterations are superfluous. To quote an instance: *Miltonia vexillaria* has become *Miltoniopsis vexillaria* but for registration purposes the genus remains a miltonia. The same situation is adopted in the names of species used as hybridising material and subsequent registration as parents of hybrids. Thus, although *Paphiopedilum glanduliferum* is the correct name for this orchid, the name *Paphiopedilum praestans* is its title if used in hybridising for registration purposes.

The hybrid list should appear as a new volume each year. Considering the size of the lists published each month in the *Orchid Review* and the *American Orchid Review* it would not be surprising if such a modification became necessary in the twenty-first century to keep the publication viable and save it from the unwieldy proportions of the other volumes. The computer has changed many old practices and it may be feasible to use it in this way.

Despite the questions posed in earlier paragraphs, it is impossible to see the end of a catalogue such as that pertaining to orchids. No other branch of horticulture has been as detailed, registered and indexed, despite the faults which are inherent in such procedures. The debt owed by growers and hybridists alike to those who instituted the index and kept it up to date to the end of the second century of orchid cultivation is beyond calculation.

Orchid Growing and the Hybrids

What began as a preoccupation with the native orchids of some countries in the world by a minority in wealthier nations has become a horticultural 'mega-index' which probably has no equal in any other branch of the plant kingdom. From being the preoccupation of a few it has spread through the world in a way unimaginable to those first cultivators. While the original species and genera are still sought by some growers, scarcely one of them has been able to confine interest to a single genus.

One of the peculiarities of human nature in orchid growing is the urge to diversify and the apparent inability to be satisfied with the product before us. We seek constantly to improve on nature, occasionally in ways which are totally against the principles of evolution or creation. This is exemplified by some of the cross-pollinations discussed in the chapters which follow, and I will try to follow it through all the phases of hybridising in the last 150 years.

It could be thought that all possible computations must surely have been exhausted by the 1990s, but apparently that is not so. The plant breeders come up with new combinations for colour, shape and size or even alterations to plant form. I do not intend to discuss the possibilities of 'genetic engineering' which I outlined in the *Australian Orchid Review* a few years ago; it is the prerogative of scientists to explain that. Perhaps they, too, will be unable to find a solution to such problems as the introduction of different colours to genera which do not carry genes for them, while preserving desirable characteristics already there.

Some alteration of the genetic code of hybrid orchids has been achieved by adjusting chromosome numbers from orthodox counts in the plant cells for each genus. We have bred things which are referred to as '3n', '4n', '6n' or polyploid plants, and this has affected the appearance, colour and size of flowers and the morphology of the plants. This is also the sphere of biologists and other scientists and the layman can barely begin to understand its ramifications. (Where the term ploidy is used it refers to the chromosome status of individual plant cells. This is usually determined by a 'root squash' and microscopic examination and count of the chromosomes. Each genus has a standard or regular chromosome count and this is nominated as 'diploid' or '2n'. Some irregularities are nominated '3n' or 'triploid', the usual result of cross-pollinating a 2n diploid parent with a 4n or tetraploid parent. Some 4n plants are natural, in similar fashion to a set of identical twins. Other diverse plants have uneven numbers of chromosomes and are termed polyploids. It is possible to chemically create odd-numbered chromosome counts in cells of most genera. (Further information should be sought in encyclopaedias.)

It could be asked what this has to do with orchid growing as a pastime, but it is relevant, even for the occasional grower who cultivates orchids which suit his or her environment in open conditions. The number of such growers increases each year, particularly with the more amenable orchids such as cymbidiums, perhaps the hardiest of all genera. Cymbidiums have captured the imagination of plant enthusiasts living in temperate to subtropical climates and they have become available in numbers and types which can satisfy the most ardent of collectors. In fact, these orchids, more than any others of the large orchid family, have become almost throw-away plants which are bought for their beauty and the lasting qualities of the flowers and discarded after some weeks when the flowers fade. They are less expensive than some flower arrangements which do not last long.

However, it is of little use asking newcomers to orchid growing to understand a single genus such as cymbidiums; a much broader view must be presented. Of the total of some 1200 genera currently named, about 600, or perhaps more, have been subjected to cross-pollination. When reduced to species level the figure becomes unbelievable. Only those cross-pollinations which have flowered and been processed through the naming system of the Royal Horticultural Society in London are significant. It is possible that these registrations comprise barely half of the real total. It must be recognised that many have been discarded as unworthy of registration and, as we all know, many hybridists are too negligent about this important stage of recognition. They do not fulfil what should be regarded as an obligation.

Since writing the first book in this series, Growing Orchids Book One, *Cymbidiums and Slippers* (the last named correctly known as paphiopedilums), I have constantly aimed to make the pastime as easy as possible to understand. I have extended this to the numbers of plants involved and how they came into existence and will take it a step further. As far as possible, the origins of hybrids will be traced through their development into the plants and flowers we look at some 150 years after the first cross-pollinations. It is not possible to completely cover all genera and those selected are models for others.

It is not always easy to trace the development of the breeders' or hybridisers' work or to determine what stimulated it. It becomes at times a matter of looking through their eyes and trying to imagine what they saw in various flowers and the combinations into which they wove them. Several factors are obvious, such as choice through morphological resemblance when combining various genera. Cattleyas, for example, are related to many other genera growing in the Americas.

The Americas and Indo-Asia are almost completely insular and the genera do not mix to any degree, except for those endemic to many countries, like habenarias, which are terrestrial and of no commercial value.

Cross-pollinating the orchids of those spheres, although frequently and ceaselessly tried, has had little or no success. This is one of the peculiarities of the orchid family as a whole. The reason for this becomes clear when we consider other plants and animals in the context of the changes in continental formations. If they were ever related it must have been in the remotest past. As in most things, there are rare exceptions, which could be taken to prove the rule.

While this detail may seem unimportant, one needs to understand it in its simplicity in order to appreciate what affects cross-pollinations in the late years of this century. Occasionally a breakthrough occurs, but these are never easy to effect and too frequently the results indicate that considerable time has been wasted. At least one instance of this is illustrated on page 73. While some cross-pollinations grow and flower easily and quickly, the general pattern is a lapse of some years between pollination and flowering.

Some tribes contain few members. One such is the Cypripediloideae, which contains three principal genera — the cypripediums, paphiopedilums and phragmipediums. Since the first member was found and named, and as others followed, many efforts have been made to cross-pollinate them.

This group of three genera was selected because it has representatives in both Indo-Asian and American spheres and cypripediums, while solely Northern Hemisphere based, are common to both spheres.

Despite the broad resemblance of the plants and flowers, they are almost incompatible. Cypripediums are terrestrial orchids. Paphiopedilums and phragmipediums are partly epiphytic, with a few species epiphytic or lithophytic as well as terrestrial. Despite the similarity of the plants, this is not a regulating factor. It goes much deeper than that.

While one of the earliest hybridisers in Veitchs' nursery succeeded in cross-pollinating paphiopedilums and phragmipediums and raised plants from the subsequent seed, in nearly all instances the flowering was either long delayed — often by years — or never occurred. In the long term it was not worthwhile because the flowers had little to recommend them.

The plants' reluctance to flower did not deter others from trying cross-pollinations. In most instances the results failed to achieve commercial recognition and the attempts lapsed. After all, it is the 'driver' of commercial exploitation which lies behind most orchid hybridising. The yearbook of registrations for 1981 to 1985 contains the only instance of naming a cross-pollination between paphiopedilums and phragmipediums. That is significant, even if it does not indicate the number of attempts.

As for the genus cypripedium itself, the members are difficult to cultivate, the numbers remaining in the wild dwindle each year and commercial or amateur growing prospects are so difficult that they have entered into orchid growing in only minor roles.

There is sufficient material in paphiopedilums and phragmipediums to keep hybridists busy even after the lapse of almost 150 years of cross-pollination. One feature is the return after many years to primary hybrid cross-pollinations, using species discovered late in the twentieth century as well as older species. These primary cross-pollinations again introduce hybrids which closely resemble those of the closing years of the nineteenth century. This factor is little in evidence in other genera, such as dendrobiums or cattleyas. In terms of appreciation, those flowers which were once derided for not having 'shape' appear to have developed a niche from which they will be hard to dislodge. Some idea of the type is illustrated on page 84. In a free survey of the development of the genus as a whole, past hybrid history indicates that time was given to cross-pollinations which led to a rounder flower than the new species are capable of producing. This, of course, leads to the view that orchid growers are fickle and not prepared to rest with past hybrid achievements. They follow 'fashion'.

It should not be thought that this trend is confined to cypripediloideae. While hybridising has satisfied certain preconceived notions of what a 'real' cymbidium should look like, the development and elaboration of the smaller cymbidiums is another facet of the overall trend. The question then arises: Are we ever to reach hybrid stability, or must similar periodic changes be seen as part of the cult?

The answer to that question involves retracing the long path over the years from the first cross-pollinations to the plants and flowers of the end of the twentieth century. It cannot all be covered in this book, but some notion of the desires, ambitions, and motives of men and women and involvements of various genera can be woven into a pattern that has been one of constant repetition. By the 1990s it appears that we are treading paths which have been trodden so many times before. The innovations have become lost in a series of permutations which are reflected in *Sander's List of Orchid Hybrids* and which number hundreds of thousands.

There is no time to dissect these entirely, but they can be analysed in a sufficient sample to give understanding of the complexity to anyone wanting to find out where it all began. The person responsible for the list gave us an ordered reference system covering all the genera used in the period of orchid cultivation, as far as was possible considering the material. This list is perhaps unequalled by any other horticultural catalogue and may be rivalled only by the list of bloodstock horses. As this book was being compiled in 1990, seven volumes of *Sander's List of Orchid Hybrids* had been published and another was due. Each volume contains the registrations for five or more years and in one instance it was necessary to divide the book into two sections because of its size. The 1981–1985 list contains 802 pages of registrations, which included more than 10 000 new names.

Frederick Sander, the compiler of the first volume, began his foreword by stating:

> The nomenclature of Orchids is a subject in which I have for many years taken a great interest. So long as our records were limited to species and varieties of species, no great difficulty arose. Wrong names may have been given in the hurry of the moment, but these were in a short time discovered and put right . . .

It is incontrovertible that this pattern of right and wrong has persisted throughout the subsequent volumes, sometimes inadvertently, sometimes deliberately. These errors are continuously being corrected, and thorough hybridists are faced with a list of corrections in each new volume or in the various publications which print the lists supplied by the Royal Horticultural Society.

The almost unbridgeable gap mentioned in the introduction occurs each time taxonomists or botanists change the names of genera or species or break down known genera into new taxa. If these genera or species are included in the first volume or any subsequent volumes as pollen or seed parents they could only be renamed with difficulty and complete reprinting of records.

As a random example, *Vanda sanderiana* appeared as the pollen parent with *Vanda coerulea* as the seed parent of the primary hybrid *Vanda* Rothschildiana in the original volume of 1945. This hybrid was subsequently used to produce further hybrids, both within the genus and with other genera. It can be appreciated that, having once become involved, the series had to be regarded as valid even after *Vanda sanderiana* was subsequently renamed *Euanthe sanderiana* by Schlechter in 1914 (and correctly so). It was very many years, however, before its new generic title was accepted.

Many genera which appear in *Sander's List of Orchid Hybrids* have been renamed and the rule applies equally to them when they have been used in earlier volumes of the list as pollen or seed parents. Each time the species appears it is followed by a short explanation that the original name:

> is retained as the horticulturally recommended name for registration purposes, even though the applicable genus and species name (in this instance *Euanthe sanderiana*) is the botanically correct name for this species.

The first *Sander's List of Hybrid Orchids*, published in 1945, used twenty-eight genera to produce the named cross-pollinations; this number had been increased to 144 by the time the 1985 edition was published. Later in the 1980s, many other genera became subjects for experimental cross-pollinations. The number of attempted cross-pollinations, which include some bizarre miscegenations, is beyond belief. Some of these, from earlier periods, are included in this book.

Although cultivar or varietal names were included in some of the earliest registrations the requirement was never compulsory and the habit was soon dropped altogether. An attempt was made to have it reintroduced in the 1980s, but it remained only a suggestion and was not implemented. This was fortunate, because it is not difficult to see that the idea was not really feasible, given the almost unwieldy size of *Sander's List of Orchid Hybrids* and the number of volumes.

Although the number of registrants in the first edition was relatively small, at 280 odd, it has increased dramatically and is likely to continue to grow. There are forty-five pages of names in the 1981-1985 volume and each page contains about sixty registrants. They are from almost every country in the world, but are dominated by the USA.

It is almost impossible to detail the number of genera and species used or the combinations of genera. These combinations usually fall into related groups, very few of which could be designated successful cross-pollinations with totally unrelated genera. In general these appeared to inherit characteristics from only one parent.

Such results give rise to speculation that the cross-pollinations had been careless or that stimulation of the seed-bearing parent occurred without the necessary exchange and sorting out of the chromosome content. I have always been inclined to favour the carelessness angle, as very few incompatible genera cross-pollinate. One of these suspect cross-pollinations is illustrated on page 125.

Frederick Sander was not the first man to compile a list of names for the hybrids raised in the early years of cross-pollination, but one must admire his persistence in what was a thankless task, particularly since it meant tracing entries from American and European orchid growers — and this included Russia. The rest of the world has never known much about the Russian part, even in modern orchid culture, but their culture persists.

An extract detailing the intergeneric cross-pollinations from 1986 to 1989 appears at the end of this chapter and is informative about this type of hybridising. The combinations had never previously been used. Once a combination of genera is registered all future combinations of these genera, in whatever order, follow the name originally given.

It is almost impossible to know what occurred before the epoch-making cross-pollination of *Calanthe masuca* and *Calanthe furcata* by James Dominy when he worked for Veitchs' nursery in 1856. The resulting seedlings were given the hybrid name of *Calanthe* Dominii and the reaction of botanists and others is described in Growing Orchids Book One, *Cymbidiums and Slippers*.

Some people were hostile to cross-pollinations of any sort. Among them was James Bateman, who was notable for growing only species orchids throughout his entire life. From the first successful attempt of James Dominy to the present day there has been a succession of generations which at times extends to thirteen and more cross-pollinations aimed at achieving certain characteristics in the flowers. It was not necessarily aimless persistence simply for the sake of seeing what would happen although, naturally, even this approach has a large part in the programme. These cross-pollinations may be followed in some of the illustrations.

As remarked earlier, the late twentieth century has been notable for a return to primary hybrids, this time using better species which have been raised by horticultural cross-pollinations and by selecting the best of their flowers for further work. In effect this follows a pattern outlined in the graph on page 55 of Growing Orchids Book Two, *The Cattleyas and Other Epiphytes*. The process has been aimed not only at achieving more symmetrical shape but also intensified colour. In total it was a successful departure from intergeneric and interspecies hybridising and perhaps created a new outlet for commercial plant raisers.

However, we are outstripping the story by introducing this topic and it would be best to return to the hybridisers and cross-pollinations of the years immediately following James Dominy's success. It should be reiterated, too, that James Dominy did not think it all up himself, but was spurred into action by the thoughts of another orchid fancier.

Of all flowers and plants, orchids have possibly the most sustained part in amateur horticulture, perhaps equalled only by life-sustaining plants such as vegetables and grains. The attraction of exotic plants and flowers is hard to explain and orchids were only an after-thought to the ferns, bromeliads, begonias and other plants which preceded them. The orchids attained equality and then surpassed these plants, probably more because of the beauty and strangeness of their flowers than for any other reason. Whatever it was, the attraction has remained.

In the early 1880s, Lewis Castle, at one time attached to the Royal Botanic Gardens at Kew, England, produced a small book titled *Orchids — Their Structure, History and Culture*. His first chapter related to fashion and stated: 'Fashion is proverbially variable, and its dictates always influence a large proportion of the community.' This is a pertinent observation when considering the subject of his book and extending that consideration to the series, Growing Orchids. He also wrote on cacti, but I only mention this in order to show his wide interest in plants.

As an orchid grower of more than fifty years' standing I am convinced of the truth of Castle's statement. Not all growers are affected by changes of fashion, but it is true that many growers fall into what could be seen as the prime error in orchid growing — abandoning what one grows well to take up another genus which comes into fashion.

Fashion also influences the changing appreciation of orchid flowers, sometimes by those who do not grow the particular genus involved and perhaps do not even understand the route by which its hybrids were produced and the genera from which they were bred. The importance of this kind of understanding cannot be too highly stressed.

Use of the word 'genus' may be misleading but, to put it into context with fashion, one should understand that the word 'cattleya' is used in its conglomerate sense far more frequently than when used to designate the genus in its pure form.

Lewis Castle, fortunately for posterity, listed the principal genera and species in cultivation at the period in which he was writing his all-too-short book. This list covered all the hybrids then grown and those which had been created in the years during which he was interested in orchid growing. The history, brief though it is, goes back to the original cultivators then known in Britain. European growers were mentioned only briefly, but were equally numerous.

Two hybridists who worked for the Veitch nursery in the 1850s appeared to dominate the cross-pollinating field. They were James Dominy, whose fame as the first of so many thousands of followers was established in 1856 with the flowering of *Calanthe* Dominii, and another employee of that nursery, Mr Seden. (The latter was referred to as 'Mr' Seden, whilst his colleague was known as James Dominy.)

Of the two men, Mr Seden had the longest list of flowering hybrids to his name for the following thirty odd years. These two men extended their work into fifteen genera, with the first inter-generic cross-pollinations included in the total list. It may be interesting to readers to know the names of the genera involved in their work and they are included below.

It should not be thought that these two men from the Veitch nursery were alone in their innovations. The trend toward this departure in orchid cultivation spread rapidly and from it emerged a plethora of cross-pollinations, many involving what was or should be termed miscegenation, using various genera. Some of these strange attempts at hybridising are included in other parts of this book.

When some of the results of manual cross-pollinations flowered it became obvious that the products were already known as species. This led to the realisation that nature had long been at work and to the inclusion of such natural hybrids in the botanical lists of orchid plants. It also caused both qualified and unqualified botanists to nominate hypothetical parents for other flowers which were in cultivation and finally to prove or disprove their origins by manual cross-pollinations.

James Dominy's Hybrids*

Aerides hybridum (*A. affine* × *A. fieldingi*)
Anoectochilus Dominii (*A. xanthophyllus* × *Goodyera discolor*)
Calanthe Dominii (*C. masuca* × *C. furcata*). (Credited as the first artificial cross-pollination raised and flowered.)
Cattleya Brabantiae (*C. loddigesii* × *C. acklandiae*).
Devoniensis (*C. crispa* × *C. guttata*) — the first-named now known as a laelia.
Dominii (*C. maxima* × *C. amethystina*)
Exoniensis (*C. mossiae* × *Laelia purpurata*)
Felix (*L. crispa* × *C. regnelii*)
Hybrida (*C. granulosa* × *C. harrisoniae*) (the first man-made cattleya to flower, so far as is known)
Manglesi (*C. mossiae* × *C. loddigesii*)
Pilcheri (*L. crispa* × *L. perrinii*)
Quinquecolor (*C. acklandiae* × *C. forbesii*)
Sidniana (*L. crispa* × *C. granulosa*)
Cypripedium Dominii (*C. piercei* × *C. caudatum*)
Harrisianum (*C. barbatum* × *C. villosum*)
Vexillarium (*C. barbatum* × *C. fairrieanum*)

(The genus phragmipedium was included here, as will be obvious in the first hybrid.)

Dendrobium Dominii (*D. nobile* × *D. moniliforme*)
Goodyera Veitchi (*Goodyera discolor* × *Anoectochilus veitchi*)
Laelia Veitchiana (*Cattleya labiata* × *C. crispa*) [illustrating the confused state of nomenclature]

*Based on Lewis Castle, 'Orchids, Their Structure, History and Culture', *Journal of Horticulture*, London, 1887.

Phajus Irrorata (*P. grandifolius* × *Calanthe vestita*)
Inquilinus (*P. vestitus* × *Calanthe masuca*) [as in Lewis Castle's book — confusion again.]

(It should be noted that all the above were catalogued as cattleyas, whereas in fact many were laeliocattleyas or pure laelias.)

This list, published in 1887, represents almost a life's work for James Dominy. Although his first cross-pollinations were possibly in the cattleya group, which included many laelias at that time, plants resulting from sowing the seed were slower to flower than the calanthe.

The seed-raising technique included the wasteful process of scattering the contents of ripened capsules on the surface of potting material around healthy, growing plants of the same genus. It was not much slower than the asymbiotic process of the early scientific years. Although this scattering gave indications that there were advantages in using pots containing plants of similar genus to the seed used, the true facts about the assistance of fungi were not even the subject of speculation. This realisation came some fifty years later when the culture of fungi and their use entered seed-raising techniques. Further information on this appeared in the *Orchid Review* of 1906, pages 201 and 203.

Mr Seden's Hybrids*

The list of cross-pollinations attributed to this hybridist is much longer than that of James Dominy. In each instance, however, the registrant is the Veitch nursery, which in some ways transfers the honour from the men to the firm which employed them. In some instances Seden's work duplicated that of Dominy. A selection of his cross-pollinations is:

Cattleya Amesiana (*C. crispa* × *C. maxima*)
Chamberlaini (*C. leopoldi* × *C. dowiana*)
Fausta (*C. loddigesi* × *C. exoniensis*)
Mastersoniae (*C. loddigesi* × *C. labiata*)
Mardelli (*C. speciosissima* × *C.* (?) *devoniana*) — (first-named is *C. lueddemanniana*).
Porphyrophlebia (*C. intermedia* × *C. superba*) — (last-named is *C. violacea*)
Triophthalma (*C. superba* × *C. exoniensis*)
Suavior (*C. mendeli* × *C. intermedia*)
Veitchiana (*C. crispa* × *C. labiata*)
Chysis Chelsonii (*C. bractescens* × *C. aurea*)
Sedeni (*C. limminghi* × *C. bractescens*)
Cypripedium Albo-purpureum (*C. schlimii* × *C. Dominii*)
Calanthum (*C. biflorum* × *C. lowi*)
Calurum (*C. longifolium* × *C. Sedeni*)
Cardinale (*C. Sedeni* × *C. schlimii*)
Euryandrum (*C. barbatum* × *C. stonei*)

*Based on Castle, op. cit.

Germinyanum (*C. villosum* × *C. hirsutissimum*)
Grande (*C. roezlii* × *C. caudatum*)
Leeanum Superbum (*C. insigne* var. *maulei* × *C. spicerianum*).

(This is an instance of the commonly included varietal names, later abandoned.)

Marmoraphyllum (*C. hookerae* × *C. barbatum*) 1876
Marshallianum (*C. venustum* var. *pardinum* × *C. concolor*)
Microchilum (*C. niveum* × *C. druryi*)
Morganiae (*C. Veitchii* × *C. stonei*)
Nitens (*C. villosum* × *C. insigne*)
Oenanthum (*C. Harrisianum* × *C. insigne*)
Porphyreum (*C. roezlii* × *C. schlimii*)
Porphyrochlamis (*C. biflorum* × *C. hirsutissimum*)
Cypripedium Porphyrospilum (*C. lowi* × *C. hookerae*)
Pycnopterum (*C. venustum* × *C. lowi*)
Schroederae (*C. caudatum* × *C. sedeni*)
Sedeni (*C. longifolium* × *C. schlimi*)
Seligerum & majus (*C. barbatum* × *C. laevigatum*)
Superciliare (*C. barbatum* × *C. Veitchi*)
Tesselatum (*C. barbatum* × *C. concolor*)
Vernixium (*C. argus* × *C. villosum*)
Winnianum (*C. villosum* × *C. druryi*)
Dendrobium Endocharis (*D. japonicum* × *D. heterocarpum*)
Euosmum (*D. endocharis* × *D. nobile*)
Micans (*D. wardianum* × *D. lituiflorum*)
Rhodostoma (*D. Huttoni* × *D. sanguinolentum*)
Splendidissimum (*D. aureum* × *D. nobile*)
Laelia Bella (*L. purpurata* × *Cattleya labiata*)
Callistoglossa (*L. purpurata* × *Cattleya gigas*)
Canhammiana (*Cattleya mossiae* × *Laelia purpurata*)
Flammea (*L. cinnabarina* × *L. Pilcheri*)
Philbrickiana (*C. acklandiae* × *L. elegans*)
Sedeni (*L. superba* × *L. Devoniensis*)
Masdevallia Chelsoni (*M. amabilis* × *M. Veitchi*)
Gairiana (*M. Veitchi* × *M. davisi*)
Phajus Irrorata (*P. grandifolius* × *Calanthe vestita*)
Phalaenopsis Intermedia (*P. amabilis* × *P. rosea*)
Zygopetalum Pentachromum (*Z. mackayi* × *Z. maxillare*)
Sedeni (*Z. maxillare* × *Z. mackayi*)

These are almost exact copies of the lists credited to the two men, with slight modifications or interpolations where considered necessary. The last entry indicates the type of nomenclature adopted for reverse cross-pollinations. This was later amended so that the hybrids derived either way bore the same name. The nomenclature of the species was correct for the period and has since been amended to modern taxonomy. In each list the pollen parent is the last named.

These two lists did not encompass all the hybridising which went on from the period of James Dominy's cross-pollinations. In Europe similar work was being undertaken, extending as far as St Petersburg (now Leningrad).

Considerable confusion occurred when orchid growers used the same parents and gave the seedlings different names. This ended with the establishment of priorities of names at a later stage, more particularly in the major contributions of Robert Allen Rolfe and Charles Chamberlain Hurst in *The Orchid Stud Book*, published in 1909, which is mentioned in other chapters.

More than thirty other hybrids were created and named by others in the thirty-five years between James Dominy's first hybrid calanthe and the publication of Lewis Castle's lists in 1887, using anguloa, calanthe, cattleya, cypripedium, dendrobium, masdevallia, thunia and zygopetalum species. No doubt other genera figured in the unpublished cross-pollinations by various growers.

An interesting addendum to Lewis Castle's list of the Dominy and Seden cross-pollinations is that of recognised, supposedly natural hybrids. It includes anguloas, cattleyas, coeloglossums, gymnadenias, laelias, nigritellas, odontoglossums (nineteen, the greatest number in the catalogue), orchis and phalaenopsis. It should be noted that the genera came from widespread regions, the majority from the Americas. In the following decades, up to the twentieth century, the number of natural hybrids either known or proposed exceeded the forty-four of this original table. It included few cymbidiums, the popular orchid of the twentieth century, which had not then appeared in numbers over the horizon of the orchid community's world.

Castle's use of the generic title cypripedium for the three species of the group may be confusing. Nomenclature at the time was undergoing transition from older concepts to a more modern series of definitions, most of which have survived to the end of the twentieth century.

Hybrid odontoglossums raised and flowered from manual cross-pollinations were unknown at the time Lewis Castle's book went to press, although many hybridists had attempted cross-pollinations. Some failures may have been attributable to poor pollinating technique rather than to other factors.

The Veitch nursery had a reputation for innovative ideas, but in reality its approach was no different from others current among orchid growers. The cross-pollination of *Cattleya trianae* with *Sophronitis grandiflora, Cattleya trianae* with *Brassavola digbyana* and *Cattleya intermedia* with *Sophronitis grandiflora* are examples of the miscegenation practised at the time, although miscegenation may be too harsh a word, considering their success. (The spelling and taxonomy is of the period when recorded.)

It is apparent that orchid growers' thinking, although perhaps understandable at the time, led to other extraordinary combinations. The relationship of the genera was perhaps guessed at but not generally known and disparate morphological characteristics were not considered.

The peculiar reasoning of some early hybridists is indicated by attempts to cross-pollinate such divergent genera as *Zygopetalum mackayi* and *Lycaste skinneri, Zygopetalum maxillare* and *Lycaste skinneri, Zygopetalum mackayi* with various odontoglossums and other combinations of genera as happened to flower simultaneously. This was miscegenation in the true sense of the word.

Somewhat surprisingly, *Zygopetalum mackayi* is stimulated to set fruit or seed capsules, as they are properly known, with the pollen of very many orchids. If plants result from sowings they are always *Zygopetalum mackayi.*

The habit of miscegenation still persists at the end of the twentieth century and cross-pollinations have been attempted between such things as bifrenarias and cymbidiums and odontoglossums and cymbidiums, to mention only two. Strange to say, a cross-pollination between bifrenaria and cymbidium was registered in the 1988 listings and named bifrenidium.

There is an obvious question here. Did the author ever try such things? My completely honest answer is 'No!'

The men who presided over what was one of the greatest epochs in horticultural history were principally wealthy individuals. Accompanying them was an equally gifted group of women — a small group who were perhaps considered out of place at the time.

These women collected and produced notable flowering plants and some excelled at illustrating literature of the time with beautiful lithographs in permanent, accurate colours. At no time since, even with the advantage of colour photography and computer analysis and colour matching, has their work been outdone. There were also artistically gifted men and examples of this art are included as photographs in this book.

The genesis of the certificate system of Horticultural Society awards to orchid flowers and plants occurred in 1861 when *Goodyera* Dominii, one of James Dominy's listed hybrids, was given a First Class Certificate. Although cultivated more for their foliage than flowers, the parents of the grex were *Haemaria discolor* and *Dossinia marmorata.* (A grex is all the plants from a single cross-pollination.) The recorded parents are listed as *Goodyera discolor* and *Anoectochilus lowii* in an entry published in the *Orchid Review,* Volume 1.

The Royal Horticultural Society, as the governing body is now known, issues lists each month of the plants and flowers which have received awards at the meetings of the Orchid Committee. This is a rare honour, considering the number of orchid hybrids grown, but plants or flowers have been submitted for consideration from almost every orchid-growing community in the world. Publication of new hybrids registered with the Royal Horticultural Society first appeared in the *Orchid Review* in 1922.

From all the evidence available, British hybridists monopolised the cross-pollination and seed raising of orchid genera until about 1880, when French and Belgian productions began to appear.

It is interesting to compare the list of registrants of the end of the twentieth century with that of the first volume of *Sander's List of Orchid Hybrids* in 1945, when the countries represented were England, France, Belgium, the United States of America, Java, Hawaii, Ceylon, Japan, one only from the whole of South America and none at all from Australia.

Although a calanthe was the first manual cross-pollination to flower, it is not proposed to return to the genus again in this book except in the illustration on page 124. They were covered in Growing Orchids, Book Three, *Vandas, Dendrobiums and Others*, and calanthe is not a widely grown genus compared with, say, cattleyas.

A quote from Volume 1 of the *Orchid Review*, which appeared in 1893 and is almost at its century mark, should be reprinted here. The review was edited and in part written by Robert Allen Rolfe until his death in 1921. He commented:

> Large as the number of artificial hybrids has now become, it is certain to be largely increased during the next few years, by the hundreds of unflowered seedlings now in various collections, not to mention the new crosses which are continually being made. The possibilities of improvement in various directions — in size, shape, and colour of flower, increased floriferousness, and vigour of constitution — seem almost unlimited, especially if crossing is undertaken with some definite object in view, and accompanied by judicious selection. Secondary hybrids are especially promising in this connection, as from these plants of mixed parentage some interesting developments are sure to arise, to enhance the beauty of our collections.

It is not necessary to comment on that summary of almost 100 years ago, except to remark how visionary and discerning this man was. He was a true 'orchid expert' without pretension.

Below are the additions to a list of inter-generic cross-pollinations since 1985, the number and diversity of which would have been unbelievable to James Dominy and Robert Allen Rolfe:

1986

Diaphanangis	Aerangis × Diaphananthe
Epiglottis	Epidendrum × Scaphyglottis
Erydium	Erycina × Oncidium
Gomoglosum	Gomesa × Odontoglossum
Knappara	Ascocentrum × Rhyncostylis × Vanda × Vandopsis
Lichtara	Doritis × Gastrochilus × Phalaenopsis
Luistylis	Luisia × Rhyncostylis
Mailamaiara	Cattleya × Diacrium × Laelia × Schomburgkia
Paulara	Ascocentrum × Doritis × Phalaenopsis × Renanthera
Pelastylis	Pelanthera × Rhyncostylis
Sidranara	Ascocentrum × Phalaenopsis
Sobenigraecum	Angraecum × Sobenikoffia
Tetradiacrium	Diacrium × Tetramicra

1987

Angreoniella	Angraecum × Oeoniella
Burkhardtara	Leochilus × Odontoglossum × Oncidium

Carpenterara	Baptistonia × Odontoglossum × Oncidium
Catasandra	Castasetum × Galeandra
Cleisofinetia	Cleisocentron × Neofinetia
Eurygraecum	Angraecum × Eurychone
Gastisocalpa	Gastrochilus × Luisia × Pomatacalpa
Helpilia	Helcia × Tricopilia
Leptodendron	Epidendrum × Leptotes
Lockostalia	Lockhartia × Sigmatostalix
Notylopsis	Ionopsis × Notylia
Odontopilia	Odontoglossum x Tricopilia
Vejvarutara	Broughtonia × Cattleya × Cattleyopsis
Zygodisanthus	Paradisanthus × Zygopetalum

1988

Alphonsoara	Arachnis × Vandopsis
Angranthella	Aeranthes × Angraecum × Jumellea
Barkonitis	Barkeria × Sophronitis
Bifrenidium*	Bifrenaria × Cymbidium
Colaste	Colax × Lycaste
Cymphiella*	Cymbidium × Eulophiella
Dendrogeria	Dendrobium × Flickingeria
Graphiella	Cymbidiella × Graphorchis
Johnyeeara	Brassavola × Cattleya × Laelia × Schomburgkia
Nornahamamotoara	Aerides × Rhyncostylis × Vandopsis
Sanjumeara	Aerides × Neofinetia × Rhyncostylis × Vanda
Trigolyca	Mormolyca × Trigonidium

1989

Bifreniella	Bifrenaria × Rudolfiella
Bollopetalum	Bollea × Zygopetalum
Brassioda	Brassia × Cochlioda
Cookara	Broughtonia × Cattleya × Diacrium
Deiselara	Laelia × Schomburgkia × Sophronitis
Fialara	Broughtonia × Cattleya × Laelia × Laeliopsis
Georgeblackara	Comparettia × Leochilus × Oncidium
Gohartia	Gomesa × Lockhartia
Gomettia	Comparettia × Gomesa
Lockochilettia	Comparettia × Leochilus × Lockhartia
Neoglossum	Ascoglossum × Neofinetia
Notylettia	Comparettia × Notylia
Rotoara	Bollea × Cochleanthes × Kefersteinia

*These two cross-pollinations in the 1988 listing used *Cymbidium pumila*, a miniature species from Asia. The registrant for both was Takaki or Tagaki Orchid Nursery. Further information was unavailable when this book was produced.

1990

Adacidium	Ada × Oncidium
Aspioda	Aspasia × Cochlioda
Caloarethusa	Calopogon × Arethusa
Charlieara	Rhynocostylis × Vanda × Vandopsis
Cischostalix	Cischweinfia × Pelatanthera
Cleisotheria	Cleisostoma × Pelatanthera
Klehmara	Diacrium × Laelia × Schomburgkia
Leocidpasia	Aspasia × Leochilus
Lockogochilus	Gomesa × Leochilus × Lockhartia
Monnierara	Catasetum × Cycnoches × Mormodes
Neograecum	Angraecum × Neofinetia
Orchiserapias	Orchis × Serapias
Tetracattleya	Tetramicra × Cattleya
Yonezawaara	Neofinetia × Rhyncostylis × Vanda
Zygotorea	Pescatorea × Zygopetalum

The principal aim of cross-pollination should be to improve existing species or hybrids to increase their aesthetic appeal, but many registrants in this list and a large number in preceding lists seem to have had some different visions in mind. The cross-pollinations which involve associated genera are credible, but others, such as the cross-pollination of bifrenaria and cymbidium, are too unnatural to be justified. An associate of my salad days once tried the same things, despite my unwillingness. Naturally nothing resulted because the cross-pollinations were, in the first place, of triploid cymbidium parents with bifrenaria and other genera. The shape characteristics were part of the attraction. Nevertheless, my associate did have something in mind which ultimately affected many hybrids — the idea of cymbidium meristems. He began with the wrong material and several years before meristemming became noticed in any branch of horticulture.

The 'Cattleya' Complex

Although a calanthe was the first manual cross-pollination or hybrid to flower, it may not have been the first genus in cultivation to produce a seed capsule which was sown or scattered naturally and produced plants. This happened previously, but we are considering only recorded history concerning those first hybrids which were cattleyas.

As this book is about orchid hybrids of all genera, I do not propose to quote authors for either species or genera, as in the other books in this series *The Specialist Orchid Grower* and *Expanding Your Orchid Collection*.

Throughout this recorded history there are many instances of poor 'bookkeeping' in the matter of names and even this first batch of hybrids was as poorly detailed as any other. The first cattleya hybrid had as recorded parents *Cattleya granulosa* and *Cattleya harrisoniae*, later amended to *C. granulosa* and *C. loddigesii*, and changed again in a third attempt to *C. guttata* and *C. intermedia*.

These Brazilian cattleya species were frequently confused in that early cultivation period, but a final analysis of their morphology, particularly that of the flowers, led to the parent plants being identified as *Cattleya guttata* and *Cattleya loddigesii*.

This first hybrid in the genus was named *Cattleya* × Hybrida and it is illustrated on page 57. In naming hybrids in the early history of orchid growing it was customary to use the symbols for male and female, indicating first the seed parent and then the pollen parent, as is common practice today.

It should be remembered that, as we have seen in James Dominy's and Seden's lists, most taxonomy was initially confused. In later pedigrees of hybrids in modern *Sander's List of Orchid Hybrids*, some names have been changed and comparative lists are advisable. Hawkes' *Encyclopaedia of Cultivated Orchids* gives some and most authoritative literature also indexes them.

The year 1859 saw the first cattleya seedlings flowering and John Lindley's remark that 'You will drive the botanists mad,' which was referred to in Growing Orchids Book Three, *Vandas, Dendrobiums and Others*, bears repetition here, if only to help us to imagine his feelings if faced with the current volumes of *Sander's List of Orchid Hybrids*.

Some botanists of the nineteenth century, appreciating the purity of the new-found race of plants which they had to name and catalogue, were implacably opposed to cross-pollination, although some came to accept it in time. One cannot but admire their ability and the thoroughness with which they carried out the task of naming and cataloguing. It was not until some years had passed that the morphological differences between cattleyas and laelias were understood. Plants of *Laelia purpurata* are still easily mistaken for cattleyas when not in flower.

Beginning with pure-bred cattleya hybrids, we will try to follow the genus through as far as necessary to bring into view the modern 'cattleya' and its potential. It may not be possible to completely cover the ramifications of all combinations, but most will be considered.

Cattleyas are usually divided into two groups, the bifoliate and the unifoliate or single-leafed species. Their use was intermixed in hybridising, particularly in the early years of cross-pollination. It is not easy to tell whether this was caused by the later appearance of single-leafed species or by the fact that the bifoliate cattleyas set seed capsules more freely. They were probably cross-pollinated both ways.

The ratio of intermixed bifoliate and single-leafed cattleyas and pure bifoliate cross-pollinations was about two to one over pure single-leaf cattleya registrations in the first forty-odd years of culture. What is known as the single-leafed *Cattleya labiata* complex was gradually introduced. The species regarded as belonging to this complex are *C. dowiana*, *C. dowiana* var. *aurea*, *C. eldorado*, *C. gaskelliana*, *C. lueddemanniana*, *C. mendelii*, *C. mossiae*, *C. percivaliania*, *C. rex*, *C. schroederae*, *C. trianaei*, *C. warneri* and *C. warscewiczii*. Other unifoliate cattleyas are bound into the hybrid register by cross-pollinations, but considering only hybrid cattleyas bred along the lines of shape and size, productions from the recognised *C. labiata* species far outnumber others from such unifoliates as *C. maxima* and *C. lawrenceana*.

All the species were cross-pollinated by about the end of the nineteenth century, but it is fairly certain that very few of their derivatives in primary hybrid form were cultivated in the late twentieth century. Like most hybrids, they have been outshone by generations of pure-bred cattleyas which carry the fashionable characteristics of shape, and there is the discard factor, prominent in early years for species and later for hybrids of all genera.

Some species selected from the thousands of plants brought into Europe had sufficient character to be retained in late twentieth century collections, but they are little used as pollen or seed parents. The pure white forms of *Cattleya labiata* varieties or species, as they are frequently called, were responsible in large measure for outstanding hybrids of that period.

In the late 1980s it was noticeable that many smaller bifoliate cattleyas, which once played only insignificant parts in hybridising, became a means of producing such brilliant, small, mixed parentage hybrids as *C.* Barbara Kirsch × *C.* Chocolate Drop. The latter is a 1965 addition to the hybrid list, derived from *C. guttata* × *C. aurantiaca*, both small bifoliate cattleyas typical of those used in 1960–1990. *Cattleya* Chocolate Drop is illustrated in Growing Orchids, *The Specialist Orchid Grower*.

It could be asked what the ultimate criteria for shape and colour in pure-bred cattleyas are. Those boundaries were probably distinguished in the middle years of the twentieth century by such flowers as *Cattleya* Princess Bells, white with a lemon-yellow disc on the interior of the labellum, which is typical of development from the albino forms of *Cattleya labiata* varieties. It is only one example of the crowning of a cattleya fancier's desires, probably equalled by clones of *Cattleya* Bow Bells or *C.* Bob Betts when well flowered.

Cattleya Bow Bells and *C.* Bob Betts are pure-bred cattleyas which came from a simple white-oriented breeding programme generated from *Cattleya* Edithiae, a secondary hybrid which has been one of the principal influences in 'cattleya' breeding. Although both were bred for albino or pure-colour flowers, the records do not make it clear whether coloured forms were also used in other cross-pollinations.

The species *Cattleya labiata* varieties used in the lead-up primary hybrid were *C. gaskelliana* var. *alba* and *C. mossiae* var. *wageneri*, which was also an albino and frequently misspelled as variety *wagneri*, even in *Sander's List of Orchid Hybrids*. But at least the varieties were recorded in those earlier years and we are able to use this to define the flowers.

The resultant primary hybrid was named *Cattleya* Suzanne Hye. The secondary hybrid bred from this line was derived from the introduction of the pollen of the *Cattleya labiata* var. *trianaei*, also the pure white form, as the original registration discloses.

This secondary hybrid, *Cattleya* Edithiae, was again a pure white and of very fine shape, even when compared with 1990 hybrids. It is one of the main sources of many high-quality flowers. Clones may still be in cultivation but they are unremarked in the flood of multicoloured flowers which are available worldwide. The grex was used as late as the 1985 list at least. The registrant of the hybrid *Cattleya* Suzanne Hye in 1906 was the Belgian Jules Hye de Crom, and that of *Cattleya* Edithiae in 1914 was another Belgian, Theodore Pauwels. Orchid growing had become international, extending over most of the continent of Europe and, so far as we can judge, to a lesser extent in America and Russia.

The time lapse of years from the discovery of the *Cattleya labiata* group to the full realisation of its potential is hard to understand. With the Pauwels registration it seemed that most of the period since hybridising began was spent in dilettante cross-pollinations without any clear vision of the future, but that, of course, is only surmise. At least *Cattleya* Edithiae signalled the advent of much of the twentieth-century catalogue.

If a second corroborative quality hybrid from about the same period is needed to validate these claims we need look no further than *Cattleya* Enid, derived from *Cattleya mossiae* and *Cattleya warscewiczii*, also known as *Cattleya gigas*. The pollen parent was again the last named and the hybrid was raised in the Veitch nursery in 1898.

Cattleya Edithiae and *Cattleya* Enid were a tiny proportion of the plants raised in what were the rose-coloured years of cattleya production. The use of the term 'rose-coloured' is deliberate because, apart from the white forms, it was the prevailing colour before the vogue for white cattleyas and the 1980s fashion for our intensely rainbow-hued selections came into being. Blue is the only colour which has so far evaded the hybridisers.

The introduction of other genera came about through some primary hybrids referred to as 'dilettante cross-pollinations' and are subjects of another section. But before leaving the cattleyas we should again consider the role of fashion in the development of *Cattleya* Bob Betts and *Cattleya* Bow Bells in the early 1950s. Both were line-bred white forms from *Cattleya* Suzanne Hye and *Cattleya* Edithiae, with shape characteristics intensified, the demand for which temporarily eclipsed that for coloured forms. The two derivatives were used with all colours, however, and combined into complex hybrids with other genera.

Brassocattleyas

The involvement of the genus rhyncolaelia in hybrid lists makes it necessary to consider the work of taxonomists or botanists in changing names to suit priorities. It should be recognised that synonymy would be a much simpler approach — rhyncolaelias should be known as brassavolas and rhyncolaelia used as a synonym. This is something which I have consistently advocated for more than a quarter of a century in various books and articles. In *Orchids Australia* in 1989 it was suggested that nomenclature should be stabilised and that some decisions already made should be reversed. This is a welcome departure from the tendency to accept the edicts of taxonomists, however well meaning, on matters that ought to be validated by world conferences before they are put into practice.

The first use of 'brassavolas', meaning the two species which have been renamed rhyncolaelias, occurred about the end of the nineteenth century. There were few tangible results, but they did not represent all the cross-pollinations. The species *Brassavola* (*Rhyncolaelia*) *digbyana* may be used as a pollen parent fairly simply, but when it is used as a seed producer the fact that the ovary is below a 'neck' which may be some centimetres long means that it may be too far for some pollens to reach the ovary.

Of thirteen early cross-pollinations of *Brassavola digbyana*, the pollen parent in every instance was a cattleya. Reverse pollinations were not recorded, but it is possible they were also carried out. Every variety of the *Cattleya labiata* complex was used, with *Cattleya schroederae* and *Cattleya* × Hardyana as outsiders. *Cattleya schroederae* is regarded by some as a variety of *Cattleya trianaei*. The first recording was in 1902. Considering that *Rhyncolaelia digbyana* was introduced to cultivation more than fifty years earlier, it is surprising that the natural curiosity of hybridists was not stimulated earlier. It was first registered as *Laelia* Digbyano-mossiae in 1889 by the Veitch nursery, recatalogued as *Brassocattleya* Veitchii in 1902 and transposed again to *Brassocattleya* Digbyano-mossiae in Sander's first list.

The brassocattleya section of the group was never as large as others and the reasons are perhaps twofold: lack of 'shape' and poor colour, unlike the laeliocattleyas and brassolaeliocattleyas, as the other combinations were named.

After removal of the two brassavolas from the genus by Schlechter, *Rhyncolaelia digbyana* took on a major role in 'cattleya' breeding or cross-pollinations and the smaller *Rhyncolaelia glauca* had little part in them.

The outstanding characteristic of the *Rhyncolaelia digbyana* flower is the beautiful fringed labellum, but in the long breeding tables this has largely become a frilled edge which at times distorts the labellum. The rather drab colour of the petals and sepals comes into prominence unless hybridising is carefully considered, particularly in earlier background cross-pollinations, and very few hybridisers have gone back to *Rhyncolaelia digbyana* in the 1980s and 90s.

It should not be inferred from the last sentence that the potential or influence of *Rhyncolaelia digbyana* was dissipated quickly. From only one of its early hybrids, *Brassocattleya* Mrs J. Leemann, bred from *Brassavola digbyana* and *Cattleya dowiana* and raised and named by Maron and Sons in France in 1902, more than 130 hybrids were generated and these in turn were extended into hundreds of other hybrids, using cattleyas, laeliocattleyas, sophronitis and sophrolaeliocattleyas in the following forty-odd years. A photograph of this old brassocattleya was used as the frontispiece to Growing Orchids Book Two, *The Cattleyas and Other Epiphytes*. I grew *Brassocattleya* Mrs J. Leemann in the late 1940s when it was difficult to buy hybrids in Australia.

It has always been a matter for regret that the two forms of *Cattleya dowiana* were not separated from the time of their discovery. One occurred in Costa Rica and was correctly noted as *Cattleya dowiana*, whilst the other originated in Colombia and was known as the variety *aurea*. It is uncertain which clone was used in many hybrids and, despite the remarkable difference in colouring, not even in the hybrids so well known in the first years of cross-pollinating was identification certain. Some of the more discerning authorities, such as Robert Allen Rolfe, could detect the differences, but their guesses were insufficient to alter the registrations. Together with other book-keeping deficiencies and mistaken entries their separate parts in all 'cattleya' breeding is unclear.

In the late twentieth century cross-pollinations for brassocattleyas are outnumbered by other combinations and where the 'end of the road' will be is difficult to guess. It should not be before the species *Brassavola* (*Rhyncolaelia*) *digbyana* is returned to the field and used again to bring back some desirable characteristics, such as the fringed labellum, which have disappeared.

The labellums in most of the brassocattleyas of 1970–1990 show an unwelcome, though not unexpected, bifurcation. Whilst some regard this as a distortion of the flowers, the labellums of most brassocattleyas have become so convoluted that they must either fold or appear with the lower lobe creased or split. In watching development of the buds it should be obvious to growers that in breeding wider labellums into flowers the folding necessary in immature buds must be intensified to fit it all in and allow for opening and maturing.

In the flower illustrated on page 62, which is several generations from species level, it should be noted how this alteration of the original labellums of both species, although enhancing the appearance of the flower, gives little chance of flat, open contours. In many cases hybridisers seem to have lost sight of this desirable characteristic.

Growers generally favour cross-pollinations which are commercially useful. Following selection of some hybrids generated for this reason, the future of brassocattleyas lies in cloning the best and it is possible to visualise further use for *Brassavola digbyana* and pure-bred cattleyas.

A glance at the list of intergeneric cross-pollinating prospects at the end of the laeliocattleya section of the 'cattleya' complex may open a window on the future for observant cross-pollinators.

Brassolaeliocattleyas

This section of the 'cattleya' complex is an amalgam of three genera in a series embodying all the best features of the hybrids as they appear at the end of the twentieth century. The first registration was in 1891 by Sir Frederick Wigan, of London, as *Brassocatlaelia* Wiganii (*Brassavola digbyana* × *Laeliocattleya* Elegans). *Lc.* Elegans is a natural hybrid between *Cattleya leopoldii* and *Laelia purpurata*. This entry appeared in *The Orchid Stud Book*, by Rolfe and Hurst.

The Orchid Review reports that this entry was given a First Class Certificate at the Temple Show in London in 1901 and gives its parentage as *Laeliocattleya* Aphrodite × *Brassavola digbyana*, with the brassavola indicated as the pollen parent. The hybrid appears in *Sander's List of Orchid Hybrids* as *Brassolaeliocattleya* Edgar Wigan.

The second registration of *Brassocatlaelia* Mackayi (*Brassavola digbyana* × *Laeliocattleya* Elegans), as it appeared in *The Orchid Stud Book*, was in 1903. This natural laeliocattleya hybrid was apparently used in both first and second registrations. The registrant was Joseph Chamberlain, of Birmingham, England, whose name was almost synonymous with the word 'orchid.' Whether he was the originator of the hybrid is debatable, but it is most probable that his cultivator, E. Cooper, made the cross-pollination, with the brassavola as seed-carrier. Cooper was also a well-known name in orchid-growing and hybridising in the late nineteenth and early twentieth centuries.

The date of recombination of the names into the modern form brasso-laeliocattleya is immaterial and there was no known significance about the order in which they were combined except euphony.

As there is now a distinction between brassavolas and rhyncolaelias, some confusion could exist if *Brassocattleya* × Lindleyana (*Brassavola tubercuata* in a natural cross-pollination with *Cattleya intermedia*) is included with this section. At the time the first records were made *Brassocatlaelia* Lawrencei, generated from *Brassocattleya* × Lindleyana and *Laeliocattleya* × Elegans and registered by Lawrence in 1897, should have been excluded. This indicates the need to keep an open mind on taxonomic changes which affect hybridism.

The reason for using the species *Brassavola digbyana* as pollinator or capsule carrier is given more fully in Growing Orchids Book Two, *The Cattleyas and Other Epiphytes*, which details the 'cattleya' complex.

No members of this intergeneric combination were listed before the 1900s, but it was only a short time until the list increased considerably. The compatibility of the three genera was never certain and occasionally it may have been a matter of finding the correct combinations to get results.

The outstanding features of the three were not completely present in each genus, with the result that for influence on the labellum shape and type the cattleyas and brassavola (in modern taxonomy rhyncolaelia) each played a part, leaving much of the crispness and shape to the influence of laelias. Of the latter *Laelia purpurata* and *Laelia tenebrosa* contributed the overall size and *Laelia pumila* — or *sincorana* as we know it — gave the crisp, better shaped petals and sepals. *Laelia pumila* or *sincorana* also contributed a better profile for the flowers, overriding the 'floppy' look of the petals of most cattleya species.

At no time in the history of hybridising has the production of any 'cattleya' combination exceeded the numbers reached by laeliocattleyas. It is hard to see any reason for this disproportionate result except perhaps the difficulty in effecting cross-pollinations because of ploidy (see pages 1–2 for an explanation of this word) or the natural barriers imposed by intergeneric combinations. Laeliocattleyas are the base material for extending the complex into brassolaeliocattleyas. A selection along certain lines, principally 'shape' (by which is implied a rounded outline and a fairly flat profile), has been the guiding code, maintained for generation after generation. One could ask, after looking at some of the publications of the end of the twentieth century, whether there is any potential left for growers to exploit.

On the question of colour, there is no more pronounced diversity than in the three possible laelia combinations with cattleyas, nor does size enter the calculations. In the last half of the twentieth century the introduction of other laelias, apart from the regularly used *Laelia purpurata*, *Laelia pumila* and the tan-to-orange *Laelia tenebrosa*, brought smaller flowers.

Smaller cattleyas, including the yellow-to-gold *Cattleya aurantiaca*, had the same effect but added to the existing spectrum. The introduction of pollen from 'blue' forms was also tried, but intensification of colour was apparently impossible. Some change in the process of cross-pollination and a little 'genetic engineering' appear to be solutions for achieving blue.

Hybridisers were very fortunate in the material they had at their disposal but, as with other intergeneric combinations, the hurdles of overproduction, underselling of complete stocks and the subsequent discarding of remainders meant that a great amount of material was left unused, untested and lost to further processing.

In the late years of the twentieth century , there was scope for increasing the numbers of intergeneric combinations, but this had the effect of rendering orthodox cattleyas, laeliocattleyas and the other members of this natural intergeneric group obsolescent in the same way as those early and middle twentieth-century hybrids. Some, however, have almost assured continuity in culture. But what of those which are at best 'just another orchid'?

It should not be forgotten in the 1990s that the work of the innovators is becoming lost and it was difficult to find reproductions, even of the best of their work, to include in this book. Most of the primary hybrids, unlike those of the 'new' producers' forms of older paphiopedilums, are almost non-existent. And perhaps the real question should be, 'Who wants them, anyway?'

Laeliocattleyas

The laelias of Dominy's and Seden's cross-pollinations are not all members of the genus, according to twentieth-century thinking. Some of the taxonomic confusion of the nineteenth century has been clarified, although it is common to note some reclassified species still bearing schomburgkia name tags.

Laelias, principally the Brazilian species, play a role in cross-pollinations which form the major part of the laeliocattleya portion of the complex. In the first volume of *Sander's List of Orchid Hybrids*, which recorded all the registered cross-pollinations up to 1945, there are about 2800 entries for laeliocattleyas. Although these hybrids have been outnumbered in later volumes, they still represent one of the largest sections overall.

An analysis of the members of the genus laelia shows that the greatest number of involvements used three or four species. This is not reflected in primary hybrids, but in tracing the pedigrees of various plants over a number of generations it is rare to find any which have no infusion of *Laelia purpurata*, *Laelia pumila*, *Laelia sincorana* or *Laelia tenebrosa* in their background. Sometimes it is only once that a laelia appears, but the influence is definite and long lasting. Thus in the hundreds of thousands of intergeneric hybrids in the 'cattleya' complex, laelias play a substantial part.

This section of the book considers the early introduction to the complex of the four laelias mentioned above but does not follow it through to the late twentieth century except in one or two instances. It is not possible to come to any conclusions about popularity of colours, shape or plant habit, or the reasons why hybridists brought so many laeliocattleyas into the list. It is likely, however, that fashion had a hand in it, with perceptive breeders carrying out the right cross-pollinations to suit future demand — or did blind chance play the major role?

The first traceable laeliocattleya to flower, in Britain at least, was recorded as the cross-pollination of *Cattleya mossiae* and *Laelia purpurata*. It was named *Cattleya* × Exoniensis. Exhibited in 1863, it was awarded a Second Class Certificate because it was judged as underdeveloped. A year later it gained a higher award.

The parentage of this laeliocattleya was later amended to *Cattleya mossiae* × *Laelia crispa*. This was accepted, but later it was found that *Cattleya labiata* was the more probable parent because its flowering period coincided with the laelia but that of *Cattleya mossiae* did not. However, the registration in *Sander's List of Orchid Hybrids* denotes *Cattleya mossiae* as the accepted parent. This hybrid was registered in 1873 as the pollen parent of the first secondary hybrid laeliocattleya.

The seed parent was *Cattleya loddigesii* and the resultant grex was named *Laeliocattleya* Fausta. The flowers were variable but intermediate between the two and the cross-pollination is credited to Veitch.

Indecision appears to surround the difference between *Laelia pumila* and *Laelia sincorana*. Although it is certain both have been used, as in the instance of *Cattleya dowiana*, we are left in ignorance. My deduction, from various pieces of evidence, is that *Laelia sincorana* has been most used, although the registration may have indicated *Laelia pumila*. *Laelia sincorana* did not appear in Sander's list until the 1981–1985 registrations, while *Laelia pumila* goes back to the original naming of the plant. *Laelia dayana* is an unknown factor, with occasional reference suggesting that it is a variety of *Laelia pumila*.

The first use of *Laelia pumila* in conjunction with unifoliate cattleyas was a three-way tie, as they were all registered in 1893. The cross-pollinations, in alphabetical order, were: *Laeliocattleya* Clive (*Cattleya dowiana* × *Laelia pumila*), *Laeliocattleya* Cornelia (*Cattleya labiata* × *Laelia pumila*) and *Laeliocattleya* Epicasta (*Cattleya warscewiczii* × *Laelia pumila*). It would be useless trying to distinguish whether *Laelia pumila* or *Laelia sincorana* was the parent, but in each instance the capsule parent was the cattleya.

It is probably true that none of the hybridists had the slightest inkling of the enormous implications of their work for future generations of people or the plants they cultivated.

To look first at the people: what to those early cultivators was a dilettante's pastime became a worldwide activity which reached into every strata of society and evolved into business enterprises in most temperate western civilisations. The Third World, where most species of orchids originated, initially had little contact with this activity, although in some countries the local population had a minor share in a tremendous export business accounting for millions of plants.

The hybrids, numbering thousands of registrations for all those cross-pollinations worthy of naming, were divided into cultivars, sometimes referred to in older literature as varieties, and accorded points of merit which governed the price asked for them. Unfortunately, the wastefulness of this system led to the discarding of unwanted or outdated clones which gradually disappeared each year and continue to do so.

Occasionally a cross-pollination produced a number of clones which were distinguished; at other times the whole cross-pollination was assessed as too poor to proceed with and was discarded. Regardless of the worth of a clone at any given period throughout the history of hybrid cultivation, it is most unusual now to be able to locate a cross-pollination made, say, in the 1930s, although that decade saw the first of the 'ideal' laeliocattleyas emerge. As for the first cross-pollinations, it is unlikely that any one of the three 1893 hybrids could now be located.

The second secondary hybrid laeliocattleya flowered in 1889, using *Laeliocattleya* Exoniensis again, this time in combination with *Laelia pumila*. This last-named orchid therefore had the honour of initiating, with *Laelia purpurata*, the long line of laeliocattleya hybrids which followed, and *Cattleya* Exoniensis assumed its correct bigeneric name of laeliocattleya.

Several bifoliate cattleyas, with laelias, were also responsible for early primary hybrids, such as *Cattleya bicolor* in 1898 and *C. acklandiae*, *C. granulosa* and *C. harrisoniana* in 1900. Nearly all the *Cattleya labiata* group had been cross-pollinated either with each other or with laelias by the beginning of the twentieth century.

This *Cattleya labiata* complex is largely responsible for the development of laeliocattleyas and although so many of its early hybrids have been discarded in the recreation of collections, the same feature did not apply to the hybrids developed from the bifoliate section. It is still possible to locate propagations of the first of these, *Laeliocattleya* Parysatis (*Cattleya bowringiana* × *Laelia pumila*) (Veitch, 1893), and several other bifoliate primary hybrids.

Laelia tenebrosa, like most other laelias discussed in this book, is a Brazilian species. Brazil occupies almost half of the South American continent and has most of the South American laelias. Most of the genus are small, but they still play a part in the production of hybrids. It is perhaps best, however, to pass over this minor part rather than give it a disproportionate place in our discussion.

Laelia tenebrosa was never as significant as others of the trio and was first noted as the pollen parent with *Cattleya bicolor* in the primary hybrid *Laeliocattleya* La France in 1898. The hybridist was Georges Mantin, of Olivet, France. The first British exhibitor of this hybrid was Charlesworth and Co. at a meeting of the Royal Horticultural Society in 1902.

The first unifoliate cattleya cross-pollination with *Laelia tenebrosa* was raised and flowered by C. Maron and Sons, Brunoy, France, in 1899, using *Cattleya mossiae* as the other parent.

Laeliocattleya Martinetti became the registered name for the hybrid, a choice from those given by four other British and European hybridists, one of which registered the cross-pollination as using *Laelia grandis*, a synonym for *Laelia tenebrosa*.

It was common in those early years, when communication between growers and hybridists was poor, to publish registrations of the same cross-pollinations in several magazines. This was a recognised form of registration, but the sorting process and allocation of prior rights to names and final acceptance was often long delayed.

The two laeliocattleyas involved use of both morphological forms of cattleya, a bifoliate and a unifoliate. This represented the beginning of the longest section included in *Sander's List of Orchid Hybrids*. So far as I know the total number has never been computed, but a rough approximation approaches 20 000. This figure is based solely on the length of the laeliocattleya list of almost 3000 in Sander's first volume, and a rough estimate of the number in the subsequent volumes until that of 1985.

In concluding the section on laeliocattleyas, a review of the pedigree of a 1989 registration indicates the number of species used to produce *Laelio-cattleya* Hawaiian Drumbeat, the immediate parents of which are *Laeliocattleya* Drumbeat × *Laeliocattleya* Hyperion, raised by Kodama in Hawaii. This shows that *Laelia purpurata* and *Laelia pumila* were involved only once each. But the cattleya species included throughout the pedigree were *Cattleya labiata*, once; *C. dowiana*, six times; *C. gaskelliana*, once; *C. lueddemanniana*, once; *C. mendelii*, once; *C. mossiae*, once; *C. percivaliana*, once; *C. trianaei*, three times; *C. warneri*, once; *C. warscewiczii*, three times; and the natural hybrid *Cattleya* × Hardyana, twice. Only one of the so-called *Cattleyalabiata* complex, *Cattleya eldorado*, missed out.

Neither *Cattleya dowiana* nor *Cattleya* × Hardyana, which stemmed from it, both in yellow-to-gold tones, produced the coloured forms expected from them and in general lost first generation colour in the transition. It was left to later hybridists to breed it back into the 'cattleya' complex.

Sophronitis

Sophronitis is a small genus of indefinite species demarcation and should preferably be regarded as a conglomerate of varieties from different regions, with perhaps two or three distinctions at species level. The general colour is red and the shape is similar to that of cattleyas or laelias. Its compatibility with both those genera has been well demonstrated.

The species commonly used in conjunction with the 'cattleya' complex is *Sophronitis coccinea*, sometimes referred to as *Sophronitis grandiflora*, particularly in older literature. It was discovered early in the nineteenth century near Rio de Janeiro, Brazil. It is worth cultivating in its own right and its influence in hybridism with the 'cattleya' complex brought additional colour and perhaps some shape to its progeny. The flowers are generally rounded, with broad segments.

The first recorded alliance was with *Cattleya intermedia*, a Brazilian bifoliate species. The cross-pollination was made by Veitch and flowered in 1886, but apparently all the plants raised from the cross-pollination were held in Baron Schroeder's collection at The Dell in England. The hybrid was named *Sophrocattleya* Batemaniana and the photograph on page 67 was taken in Australia in 1988.

As in other instances, history has differing versions. In researching this cross-pollination another version came to light. Mr Seden made the cross-pollination, using the sophronitis as the seed parent and the hybrid flowered in 1886, after a lapse of five years from sowing. Reichenbach referred to it as 'a lovely gem, a miniature laelia', to which he added: 'This novelty offers a wide field for considerations of nomenclature. Are all hybrids between what we call genera to get intermediate names? The effect of mixing a sophronitis and a cattleya is a laelia; hence I must reduce sophronitis to laelia, except *S. violacea*, with a remodelled character.' From that it is possible to see the taxonomic strength of the botanists of the era. The two versions are not irreconcilable.

The first hybrid with a unifoliate cattleya was in combination with *Cattleya mossiae* as the seed parent. It was named *Sophrocattleya* Imperatrix. Raised by Veitch, it continued an almost unbroken line, using sophronitis as the pollen parent. This could have been deliberate, but unfortunately no records exist of reverse cross-pollinations, so it is not known if these proved nonviable.

Several sophronitis-laelia cross-pollinations were made in the last years of the nineteenth century and these survived through to the present time, some ninety years later. Two of the best known flowered in 1901–1902. They were *Sophrolaelia* Orpetii (*L. pumila* × *S. grandiflora*) and *Sophrolaelia* Psyche (*L. cinnabarina* × *S. grandiflora*). Some illustrations on page 67 are of plants from my collection.

In each instance *Sophronitis coccinea* syn. *S. grandiflora* was used as the pollen parent. Many growers throughout the world still grow and admire these two, although others which appeared about the same time seem to have disappeared from general cultivation.

Interbreeding of sophronitis, laelias and cattleyas has produced the long but easily remembered name sophrolaeliocattleya, while the addition of brassavola (rhyncolaelia) to this combination has been given the coined term 'potinara'.

There are several other combinations of genera, each of which has been given a coined name, such as 'Fergusonara', the alliance of brassavola, schomburgkia, laelia and sophronitis. The origin of the name potinara and all the other names ending in 'ara' was given in Growing Orchids Book Two, *The Cattleyas and Other Epiphytes*, page 163.

It is not proposed to include a list of all the cattleya alliances, which are given in full in the various volumes of *Sander's List of Orchid Hybrids*. The combinations with brassavola (rhyncolaelia), laelia and sophronitis are the most important in the 'cattleya' complex.

Other genera compatible with this complex include barkeria, brassavola, broughtonia, cattleyopsis, diacrium (caularthron), domingoa, epidendrum (including encyclia), laeliopsis, leptotes (tetramicra), schomburgkia and other less used but still important species and hybrid combinations, many of which could be termed miscegenation.

Some of the combinations produce smaller but more brilliant flowers, most in the red to rose-purple colours, which have had transient popularity. There was also a trend to use albino forms of all species in the late twentieth century, not just within the 'cattleya' complex.

For about twenty-five years English hybridists almost monopolised 'cattleya' complex breeding, but in 1871 the French produced *Cattleya* Calummata, the hybrid from *Cattleya intermedia* and *Cattleya acklandiae*. It came from Alfred Bleu, of Paris, who later specialised in miltonias.

The Hybridising Labyrinth

A potinara is a good example of the degree to which cross-pollinations have been taken in the 'cattleya' complex. The grex named *Potinara* Hawaiian Landmark in 1989 and illustrated on page 68 contained thirteen cattleya species and one natural hybrid cattleya, four laelia species, a brassavola (rhyncolaelia) and one sophronitis in the pedigree.

The list comprised *Cattleyas aurantiaca, bicolor, forbesii,* and *harrisoniana* with one entry each from the bifoliate cattleya section; from the unifoliates *Cattleya labiata* had one entry, *C. dowiana* eight, *C. mossiae* two, *C. rex* one, *C. schroederae* one, *C. trianaei* one, *C. warneri* one, *C. warscewiczii* two and *C.* × Hardyana three; from the laelias, *Laelia cinnabarina* had one, *L. flava* one, *L. milleri* one and *L. tenebrosa* four; from brassavolas, *B. digbyana* and *Sophronitis coccinea* each had one.

Any inferences which may be drawn from the combination of these species and the three infusions of *Cattleya* × Hardyana, the natural hybrid between *Cattleyas warscewiczii* and *dowiana*, could be at best only the wildest of guesses.

The plant could best be described as a decorative semi-miniature cluster type or a plain cluster type. This class of orchid became very popular from 1970 to 1990.

When assessing such plants we need to remember that more than five years could have elapsed between the cross-pollination and the ultimate flowering. It is possible in some instances of such diverse backgrounds that much variation could result in flowering clones, some good and some worthless.

The usual pattern with clones such as the one illustrated is to replicate them into considerable numbers of mericlones and make them a commercial proposition which is not marketed until there is a prospect of recouping outlay.

Other examples of the 'cattleya' complex show orthodox or unorthodox cross-pollinations. Some are sterile 'mules', the term which the early taxonomists and botanists originally applied to most hybrids. Sterility occurs in most genera as the breeding lines lengthen and when this happens further work is abandoned or chemical adjustments are made to mericlones to try to restore breeding potential.

Finally, before we leave the 'cattleya' complex, we should consider how far hybridising can be taken. Can we go on indefinitely searching for the rainbow or the pot of gold that hides at the end of it?

It seems probable that there is a line beyond which it is impossible to go. No doubt as human beings we are very clever — perhaps too clever by half, as someone once remarked. It all leads back to a philosophical analysis of what we are doing and why we are doing it.

Brassavolas

Occasionally every orchid grower must regret necessary nomenclature changes, particularly if they affect a genus such as this, which tends to lose its identity when the principal members have been transferred to another genus. Inevitably this also confuses the records of bygone years and complicates study of its hybrids. Brassavolas, as distinct from *Rhyncolaelias digbyana* and *glauca*, are dominant and this is borne out when their hybrids are considered.

The genus is usually thought of as allied to cattleyas and similar genera, but it is a group which has a steady background of hybridising going back to 1900, when the Brazilian species *Brassavola fragrans* syn. *B. perrinii* was used with *Cattleya intermedia* to create the hybrid *Brassocattleya* Nivalis. The flower closely resembled the natural hybrid *Brassocattleya* × Lindleyana (*Brassavola tuberculata* × *Cattleya intermedia*), also from Brazil.

Use of this Central and South American genus, excluding the two rhyncolaelia associates of the cattleya complex, was sparse over the following years, so that only two further hybrids were registered until 1907. The first was *Brassocattleya* Sanguinii (*B. fragrans* × *C. mendelii*) in 1904 and the second was *Brassocattleya* Mary (*B. nodosa* × *C. lawrenceana*) in 1907.

In the first half of the twentieth century those hybrids stemming from *Brassavola digbyana* outnumbered the total from the remainder of the genus.

In that period only one cross-pollination was named using two species, *Brassavola cucullata* × *Brassavola digbyana* (named *Brassavola* David Sander), although the various species were used in twenty-two other cross-pollinations in the cattleya complex. In modern nomenclature that was more correctly an intergeneric cross-pollination, but it is recognised under its old botanical name.

Only *Brassavola cucullata* and *Brassavola nodosa* carried through with cross-pollinations in the last half of the twentieth century and then principally as ancillaries to the brassolaeliocattleya complex. Extraneous cross-pollinations occurred with various composites in what could be termed 'exploratory' hybridism, with no particularly noteworthy results, except perhaps their extraordinary compatibility.

Brassavolas, as distinct from the rhyncolaelia component, are resilient, hard-looking plants with terete or semi-terete foliage and, thus, great resistance to bright light and heat which ordinary genera, such as cattleyas, find intolerable. They offer hybrids more suited to tropical conditions, provided they are not exposed to prolonged overwatering when dormant. They always grow in warm, humid conditions.

As a genus they have little to offer growers seeking shapely flowers which will win awards, given primarily for rounded outlines. But as plants which may be built into spectacular exhibits or grown for the owners' sheer pleasure, brassavolas have much to offer. The species' labellums are usually dominant, although perhaps light coloured and lacking in contrast. The flowers have spidery outlines similar to those imparted by brassias. Two hybrids are illustrated on page 73.

Cymbidiums

Unlike the 'cattleya' complex, in which so many of the genera are related, albeit indistinctly at times, the cymbidium genus is monotypic — there are no apparent related genera, although a cross-pollination with cymbidiella was achieved in the late twentieth century, using a miniature Asian cymbidium. It is not the only monotypic genus. Several others, such as sophronitella, are so constituted.

The hybrid members of the cymbidium genus have been born out of a group of species, all of Indo-Asian origin, extending from northern India south-east to Australia. In the beginning this south-eastern extension played little part in generating cross-pollinations, but despite paucity of numbers, the smaller species from this region have had quite an impact on the development of miniature cymbidiums.

Hybridism has gone through several phases, similar to those followed by other genera occupying places in *Sander's List of Orchid Hybrids*. The development of the hybrid structure began in ignorance, because nobody had the least idea what could be expected from cross-pollinations like those of the late twentieth century. Growers might have taken a lead from the hybridism of other genera such as the 'cattleya' complex, but the other orchids were so different that they could not satisfactorily be used as a guide.

The discovery of various cymbidium species and their uses were discussed in Growing Orchids Book One, *Cymbidiums and Slippers*. Some of that information is repeated here, however briefly, to satisfy readers who do not have the first book.

In all there are more than fifty species of cymbidiums, the nomenclature of which has frequently been amended. I find some of these name changes debatable, but taxonomic decisions are not my role. As in most hybridising, however, the decisions have been left too late to have much effect on *Sander's List of Orchid Hybrids*.

The true Indo-Asian cymbidiums from the area from northern India to Thailand, contained the genetic material used to produce excellent standard-type cymbidium hybrids. Those species appear, on cursory examination, to have inbreeding as a major part of their evolution, but modern techniques have not yet helped us to determine the effects of this inbreeding. Such an investigation, in any case, would not necessarily help hybridists much, apart from satisfying their curiosity. Hybrid development appears to have outrun the need to know.

The list of species involved in producing standard cymbidium hybrids includes *Cymbidiums eburneum, erythrostylum, giganteum, grandiflorum* (syn. *hookerianum*), *insigne, i'ansonii, lowianum, parishii* (occasionally referred to *C. eburneum*), *schroderi* (an indeterminate species), and *tracyanum*. Some authorities refer to *Cymbidium i'ansonii* as a variety of *Cymbidium lowianum*. This last named species was introduced many years before others.

Cymbidium species were initially used rather indiscriminately and it was left to later cross-pollinators to distinguish between those producing early, mid-season and late-season flowers. This is easily understood because the genus was neither grown extensively nor used in commercial cross-pollinations in the late nineteenth century.

While it is unimportant when considering cymbidium hybridism in the long term, it is noteworthy that the first cymbidium species to appear in the available literature was *Cymbidium giganteum*, which appeared in notes and an illustration in Lindley's *Sertum Orchidaceum*. The author remarked:

> The accompanying plate has been prepared after a drawing made at the time of its discovery and liberally placed at my disposal for publication by the Honourable Court of Directors of the East India Company. The plant itself may be soon expected in our gardens, if indeed it does not already exist there.

Sertum Orchidaceum was published in 1838, a large volume in its original form and illustrated with superb lithographs. The choice for twentieth-century buyers is probably limited to the reproduction which was published about 1978. Either is worth owning.

The illustration accompanying the text shows the flower raceme erect, but a glance indicates that it was originally pendant. *Cymbidium giganteum* was not the first of the genus in cultivation, however, and Growing Orchids Book One, *Cymbidiums and Slippers*, should be consulted for more information.

The first hybrid from the Indo-Asian cymbidium species flowered in 1889. It was named *Cymbidium* Veitchii in the *Gardeners' Chronicle* in 1889, bred from *Cymbidium eburneum* as the seed parent and *Cymbidium lowianum* as the pollen parent. It appears in present-day literature as *Cymbidium* Eburneo-lowianum and it is the hybrid progenitor of almost our total catalogue. The reverse pollination is incorrectly named *Cymbidium* Lowio-eburneum.

Between 1900 and 1908, *Cymbidium insigne* came to Sander's nursery at St Albans, England, from the collector Micholitz, who brought it from that part of Indo-China now known as Vietnam. The history of its discovery and that of other Indo-Asian cymbidium species is covered in Growing Orchids, Book One, *Cymbidiums and Slippers*.

Cymbidium insigne, although it is difficult to give it more prominence than other species, played a major role in the development of hybrids in the late twentieth century. It made no difference whether it was used as seed or pollen parent. It was the pollen parent when used with *Cymbidium* Eburneo-lowianum to produce *Cymbidium* Alexanderi, named after the hybridist H. G. Alexander, an orchid grower employed by Sir George Holford. Alexander, considering the importance of this one cymbidium hybrid, is immortalised in a way given to few horticulturists. The cross-pollination was named in 1911 and the credit in *Sander's List of Orchid Hybrids* went to Holford.

Other early hybrids generated from *Cymbidium insigne* include:

Cym. Doris (*Cym. insigne* × *Cym. tracyanum*), raised by McBeans in 1912, an early-flowering hybrid which had only minor importance in subsequent years.

Cym. Ceres (*Cym. insigne* × *Cym. i'ansonii*), raised by Hamilton-Smith in 1919, probably as important in background for pink and red cymbidiums as *Cym.* Alexanderi was in all colours.

Cym. Pauwelsii (*Cym. insigne* × *Cym. lowianum*), raised by the Belgian Theodore Pauwels in 1911. *Cym. lowianum* was responsible for flower numbers and length of the flower stem, a characteristic shared by *Cym. insigne* but with fewer flowers.

Cym. Albanense (*Cym. insigne* × *Cym. erythrostylum*), raised by Sander in 1915. This hybrid is one of the principal progenitors of early-flowering cymbidiums, a quality imparted by *Cym. erythrostylum.*

Another early-flowering species, *Cym. tracyanum*, had the weakness of short-lived flowers and did not receive the same attention. It also was prone to poorly coloured, streaky blooms.

It is thus apparent that the hybrids of this genus have a shorter history than those of many other genera, the slow development stemming perhaps from lack of appreciation of their potential.

There were a number of other primary hybrids, but rather than devote time to them it is best to leave them in limbo, where they ended their existence. The secondary hybrid *Cymbidium* Alexanderi, for example, produced the clone given the varietal name 'Westonbirt'. Apparently the cultivar was susceptible to virus from the start and has remained so. But as a progenitor of most quality cymbidiums its position is unchallengeable and H. G. Alexander's early recognition of its character indicates his perception.

The role of *Cymbidium* Alexanderi 'Westonbirt' as a seed parent when pollinated by *Cymbidium* Kittiwake, leading to *Cymbidium* Rosanna, was probably the most important in the hybridising history of the genus. The cause of the outstanding quality in *Cymbidium* Alexanderi 'Westonbirt' was not then understood and Alexander selected the clone for shape and size. Experience with the genus led him to breed from it. *Cymbidium* Rosanna was one of the parents of *Cymbidium* Balkis (1934), the other being C. Alexanderi 'Westonbirt' in a back-cross. A look at *Sander's List of Orchid Hybrids* of 1946–1970 justifies any extravagant claims for either grex.

The choice of *Cymbidium* Kittiwake remains a mystery, but the genesis of the tetraploid or 4n cymbidium could be said to have begun there, when it was used with *Cym.* Alexanderi. The ploidy of *Cymbidium* Kittiwake is unknown, because like so many of the earlier hybrids it has probably long been discarded. It may have been a chance tetraploid or 4n, giving it also outstanding character.

Cym. Alexanderi 'Westonbirt' is the only recorded clone which proved to be 4n in that grex, although in my experience at least two others were suspected but never finally proved to be so. One was known as McBeans variety, but it had poor flower numbers, a failing also shared by *Cym.* Alexanderi 'Westonbirt'.

In a study I described in the 1956 volume of the *Australian Orchid Review*, Dr Margaret Blackwood of the University of Melbourne School of Botany, kindly consented to collaborate with me in proving a number of cymbidiums. I had a list of many plants from which we extracted root tips and Dr Blackwood subjected them to examination. Her thorough work left little doubt that some clones had been misrepresented.

In order of merit the primary hybrid *Cymbidium* Pauwelsii closely followed *Cymbidium* Alexanderi. Unlike *Cymbidium* Alexanderi 'Westonbirt', this orchid in its initial form was robust, and the variety Comte de Hemptinne also proved to be a first-generation 4n hybrid, along with at least one other, illustrated in Growing Orchids Book One, *Cymbidiums and Slippers*.

It was not until the early 1930s that much care was taken about the quality of breeding clones and a large amount of the stock available at that time was from haphazard cross-pollinations of primary and secondary hybrids, which were quickly discarded, in Australia at least.

Although there were other first and second generation cymbidiums which had an impact on hybrid production, *Cymbidium* Alexanderi 'Westonbirt' and *Cymbidium* Pauwelsii were progenitors without equal, the first named in the field of white-and pastel-tinted flowers and the second in parti-coloured and brown-red tonings. It may be wrong not to recognise other clones in the production line, but once started on that trail it could be endless.

Hybridists demanded an adequate number of flowers, clear colours and good size and shape, and each was equally important. When a successful combination of these was achieved, in most grexes the number of plants carrying all the desirable factors was small; sometimes less than 1 per cent. Such clones were distributed slowly as propagations grown from divisions.

All this changed with the introduction of the mericloning process and it is now common to find that thousands of propagations, sometimes unfortunately sold when they are too small, are on the market fairly quickly after a clone is recognised.

Australian and American cymbidium growers profited from a dispersal of breeding clones from England and Europe in World War 2 and this was perhaps why Australia and America produced so many quality cymbidiums. It may have been the postwar marketing of Australian flowers in America which prompted many larger producers of cymbidium plants to extend their growing facilities and begin using *Cymbidium* Balkis and *Cymbidium* Pauwelsii parent stock, to produce magnificent grexes with better substance and colour. Whatever it was, the cymbidium, from being an exclusive plant, became the stock in trade of many thousands of gardeners in temperate and sub-tropical climates. It proved itself an easy plant to grow and flower with minimal care.

The green to yellow cymbidiums were developed from *Cymbidium grandiflorum*, an Indo-Asian species from northern India to Burma, cool growing and early flowering. It had as a counterpart *Cymbidium lowianum* var. *concolor*, a green without the lines which the common forms display.

The extent to which the primary hybrids in this green to yellow line have been developed is illustrated in Growing Orchids Book One, *Cymbidiums and Slippers*, together with a pedigree table for *Cymbidium* Jubilation 'Geronimo', one of the best cymbidium hybrids. It is unnecessary to re-iterate that section here.

Hybridists have taken cymbidiums to such lengths that the size of the cymbidium section in *Sander's List of Orchid Hybrids* and the general character of the flowers from 1985 onwards indicate that their potential must be almost exhausted. The fact that this development has been accomplished naturally and unnaturally can be seen in the illustration of the grex *Cymbidium* Lunagrad on page 80 in this book and in Growing Orchids Book One, *Cymbidiums and Slippers*.

Attention has turned to miniature and intermediate cymbidiums, apparently to exploit their potential to a similar degree. It will be a repetition of the process in which it has taken more than half a century of breeding to bring standard cymbidiums into the 1990s, and the results are predictable.

This genus has also been the subject of attempted miscegenation. I have been told of cross-pollinations with laelias, zygopetalums, bifrenarias, maxillarias, vandaceous orchids, odontoglossums and even paphiopedilums. Most of the pollinators were confident, too. All that wasted effort!

Cymbidium insigne was one of the 'finds' of the twentieth century and it is often a cause for wonder that the genus was neglected for South and Central American orchids in the early years of cultivation and cross-pollination. It was never easy to get it to flower — the plant usually had to be well grown and to fill the pot before this stage occurred. The pink form was principally used in cross-pollinations and as these progressed in thoughtful combinations the colour intensified.

The influence of the pink *Cymbidium insigne* var. *sanderi*, when combined with *Cymbidium i'ansonii*, brought into the hybrid range nearly all the pink and red flowers which grace the collections of the 1990s. The primary hybrid was named *Cymbidium* Ceres in 1919.

The catalyst in initial pollinations appeared to be *Cymbidium i'ansonii*. This is judged by some botanists to be a form of *Cymbidium lowianum*. In my experience as a grower of the orchid for so many years, it does not qualify on several counts. The principal of these are the different flowering times, number of flowers and the habit of the inflorescence. Perhaps I do not see things as botanists do, but my opinions were formed from those of Robert Allen Rolfe and his analyses in *The Orchid Review* and from personal experience in orchid growing. I followed the use of different forms of *Cymbidium* Ceres from its beginnings.

There is no doubt that the brilliance of orchids like *Cym.* Panama Red (illustrated on page 77) has never occurred from cross-pollinations using *Cymbidium lowianum*, even in the brightest clones of *Cym.* Pauwelsii. That may be said to prove nothing, but I believe that *Cym. i'ansonii* should be ranked as a species.

Most Australian cymbidiums came from England, with a minor part from Europe, in the years just before and during the early part of World War 2. Some came from experimental hybridising and some of the parentage was disbelieved. But a great amount of useful and beautiful material was obtained, including several forms of *Cymbidium* Ceres.

Before World War 2 cymbidiums were cultivated in small numbers by a limited group of orchid growers. When I started in the mid-1930s little was available, other than species and primary hybrids, and most of those were exchanged only with friends. Cymbidiums were very difficult to obtain. By the end of the war a large number of prewar cross-pollinations were flowering and they changed people's conceptions of cymbidiums in much the same way as occurred in the 1970s, 1980s and 1990s. Some of these flowers appear in the colour illustrations in this book.

During the years of transition to better quality flowers, from 1940 onwards, the effect of cymbidium breeding in America became obvious and hybridists and growers in climates suitable for sheltered growing became more numerous. Australian growers began to turn to American suppliers, rather than their traditional English sources, and orchid growing became more and more competitive as flowers were exported and cultural societies were formed throughout the world.

Because of the style of flower needed for competitive growing, much material grown from 1940 to 1950 was sold off. This had repercussions in bringing into the cult numbers of people who had formerly imagined orchid growing to be a pastime needing protected environments and a lot of money.

While it is difficult to analyse clones to show their make-up from species level, I have chosen illustrations which help to explain the build-up of visible characteristics traceable to those species, yet which also have hidden characteristics. Two of these are analysed below and those interested may trace a complete pedigree and compare the hybrids with their progenitors, most of which are illustrated in Growing Orchids Book One, *Cymbidiums and Slippers*.

It is inevitable that varied, complex hybrids must result from cross-pollinations over several generations of mixed parentage from a number of species. Some features remain constant, however, such as the markings on labellums, which, however much they may vary from species to species, still have traces of dominant features, such as internal or external markings.

One of the selected clones is *Cymbidium* Lunagrad, named by Graves in 1967, bred in America from the seed parent *Cym.* Miracle and the pollen parent *Cym.* San Miguel. It is thus one of the cymbidiums generated in the second wave of change which began about 1950. While the original grex was not competitive, it had characteristics which qualified it as clear coloured in lemon-green, with the prospect of clear green in correct cultural conditions for that colour. The flower shows quite clearly the effect of treatment for changing ploidy in the original, from 2n or diploid status to 4n (tetraploid) or polyploid form. Polyploid clones have uneven numbers of chromosomes.

The dissection of *Cym.* Lunagrad's pedigree shows twelve infusions of *Cymbidium eburneum*, eleven of *Cym. lowianum*, eleven of *Cym. insigne*, two of *Cym. parishii* (referred to occasionally by some authorities as *Cym. eburneum*), two of *Cym. tracyanum* and four of *Cym. grandiflorum*.

A feature was the inclusion of *Cymbidium* Doris Aurea, a copper-red derived from *Cym.* Lysander × *Cym.* Chiron in the pollen parent line. The colour of this fourth generation hybrid could have come only from *Cymbidium tracyanum*, a species with a minor role in the production line of cymbidiums generally. The inclusion of *Cymbidium* Doris Aurea means that there is a chance for colours other than green in succeeding generations even when they are pollinated for greens.

The second pedigree to analyse is that of *Cymbidium* Sensation, an Australian hybrid from *Cym.* Spartan Queen and *Cym.* Fascination, raised by Wondabah Orchids and named in 1961. It is also a product of the second wave. The hybridist once told me that the proportion of good quality clones to poor was about one to fifty, in a cross-pollination made for red cymbidiums.

The pedigree of this grex shows six infusions of *Cymbidium eburneum*, twelve of *Cym. lowianum*, eight of *Cym. insigne*, one of *Cym. parishii*, two of *Cym. tracyanum*, one of *Cym. grandiflorum*, one of *Cym. cooperi*, one of *Cym. garnet* and four of *Cym. i'ansonii*. The last three provided the colour background.

Cymbidium cooperi is known as the natural hybrid between *Cym. insigne* and *Cym. schroederi*. The latter cymbidium is part of the puzzle attached to the *Cymbidium insigne* complex and has never been satisfactorily explained. *Cymbidium garnet* is known as the natural hybrid between *Cym. lowianum* and *Cym. parishii*.

The last standard cymbidium to be surveyed for its antecedents is *Cym.* Patricia French, which honours a notable New Zealander. The grex, registered in 1990, embodies all the major species elements, as do most standard-grade cymbidiums, and this background gives it the unpredictable colour inheritance carried by most hybrids. Perhaps the inclusion of more of one particular species than others might make the results more predictable. For example, if the pink form of *Cymbidium insigne* was used this could be the base colour.

Cymbidium insigne appears in the pedigree twenty-three times, *Cym. lowianum* fifteen, *Cym. eburneum* thirteen, *Cym. i'ansonii* four, *Cym. tracyanum* twice, *Cym. grandiflorum* once and *Cym. garnet* once.

Included in the breeding of this hybrid are several 4n or tetraploid clones. It has always been debatable whether the tetraploid form of hybrid, the diploid (which is the natural form), or the polyploid is the ideal type to use in cross-pollinations. Most hybridists keep their own counsel on the subject. Some clones, in particular *Cymbidium* Wallara, supposedly a tetraploid and one of the parents of *Cymbidium* Patricia French, has probably never been released in its pure form, only in mericlones. In an analysis of its pedigree in Growing Orchids Book One, *Cymbidiums and Slippers,* I expressed doubts about the accuracy of the pedigree of *Cym*. Wallara.

I can remember when growers realised the differences between clones when selecting breeding material and began to understand these differences. The second wave in hybridising was focused principally on combining diploids (2n) and tetraploids (4n), which produced so many thousands of triploids (3n). This is the part of the cymbidium breeding cycle I refer to as the 'second wave'. Some of these flowers are illustrated. The introduction of this form occurred in the 1930s and persisted for about twenty years. As already stated, this led to the abandonment of most of the first-wave forms, one of which is illustrated on page 74.

The importance of the three phases in developing cymbidiums cannot be overstated. The second phase was generated principally by two cymbidiums, then the number increased as the phase progressed. The two were *Cymbidium* Alexanderi and Pauwelsii. Later additions included *Cymbidium* Rosanna and Balkis. Included in seedlings I raised in those early years was the back-cross between *Cymbidium* Alexanderi and Balkis, later named *Cymbidium* Joan of Arc by another grower. Like its parents, it was a tetraploid or 4n grex, although it is never certain that such progenitors will yield 4n clones.

These and many other tetraploid cymbidiums led to the third phase, which we entered in the late years of the twentieth century. For want of a better name it could be termed the tertiary phase, when most of the desired characteristics had been gained. It also led to the introduction of diverse smaller species to form diploid (2n) or triploid (3n) intermediate and miniature type cymbidium hybrids. The initial plants in this group were raised in the 1930s and information about them was included in Growing Orchids Book One, *Cymbidiums and Slippers.*

Cymbidium growers throughout the world have often found it difficult to keep up with trends, or perhaps we should say fashions. The discarding of the first wave and elements of the second wave of cymbidiums drew thousands of people into the pastime. It is interesting to note the way a market has been created over the years since this genus became so popular. It began with exhibitions of orchid plants in flower which were not always cymbidiums; but this genus created backgrounds to achieve mass effects. This might be disputed by fanciers of other genera and their hybrids, and perhaps tropical fanciers who grow phalaenopsis have an equal claim. But cymbidium plants and the quantity of flowers they produce in temperate and subtropical climates or easily contrived environments make them ideal exhibition subjects. Other genera provide the trimmings and a lot of brilliance.

These exhibitions provided sufficient drive to induce horticulturally minded people to try new plants with long-lasting flowers. It became a common practice to have sales sections at exhibitions and other venues where the surplus was preferably displayed in flower.

The miniature and intermediate cymbidiums came after this stage, when cultivation had become a popular pastime. The miniatures are still not totally bred to withstand the conditions of culture which standard cymbidiums tolerate. This is in part due to the species used to generate them, as some come from warm climates.

Some cymbidium growers admit that the miniature or intermediate plants in their collections take up more space than standard plants because many of them carry pendant flower stems. While these stems may be tied erect and can flower in that position, unless the inflorescence clears the leaves it is lost and the plant loses significance.

Species used to produce miniature and intermediate types include: *Cymbidiums bicolor, canaliculatum, dayanum, devonianum, ensifolium, floribundum* (syn. *pumilum*), *gyokuchin, madidum, mastersii, pumilum* and *suave*. When cross-pollinated with standard species or hybrids they produce miniature flowering hybrids. The morphology of the plants is dictated in varying degrees by the species and hybrid plants with which they are cross-pollinated and there is no indication that alternative use as pollen or seed parents influences this.

The separation of intermediate from miniature cymbidiums is most often determined by measurement across the face of the flower from petal tip to petal tip, with emphasis on the form of the flower.

The term 'form' is taken to mean the rounded outlines of petals and sepals sought since the craft of hybridism was initiated so many years ago. Unlike early cross-pollinations, select breeding has established this form on most cymbidium hybrids registered in the last twenty years.

The cymbidium jigsaw cannot be left without considering the part played by clones which did not conform to normal ploidy for the genus, where forty chromosomes to a cell indicates diploid or 2n plants. The method for determining this number is explained in the *Australian Orchid Review*, Vol. 21, No. 4, December, 1956. Some hybridists rely on the visual examination of leaves, but this is totally unreliable. It can best be done by a count of chromosomes under a microscope.

The original species from which the cult started were nominal 2n plants. It is probable that some of these species were tetraploids, meaning that the cells contained chromosomes with doubling of numbers to 80. These were natural tetraploids, although information that they occurred and in which species is not commonly detailed. Let us conclude that the principal clones in which we are interested were products of cross-pollination and appeared in hybrid form.

When cross-pollinated they could be expected to produce seedlings which were copies and carried eighty chromosomes. This was confirmed in our tests, but it is also possible for such grexes to produce nonconforming clones.

When cross-pollinated with 2n or diploid clones these tetraploids brought into being the triploid or 3n branch of hybrid production. This second wave referred to earlier principally comprised these plants and this line of cymbidiums changed the horticultural and competitive significance of orchid growing as a whole. The plants were more robust and the flowers were larger and had more substance. They led to the discarding of those first-wave cymbidium hybrids which did not recommend themselves as superior. In reality many were lost which should have been kept.

More than any others, *Cymbidium* Alexanderi 'Westonbirt' and *Cymbidium* Pauwelsii 'Comte de Hemptinne' were responsible for this dramatic change in appreciation. It did not end there, however, and when these two cymbidiums were cross-pollinated with each other or with other chance-found tetraploids another new class was produced. This was the third wave, the 4n breed, from which plants are now being used to create other classes of tetraploid and polyploid plants as an amalgam which has bridged all the gaps and reached the ultimate potential of the genus. Ploidy is still, however, very much a matter of guesswork.

One small point remains to be clarified. This concerns the matter of amalgamating the standard 2n, 3n and 4n or polyploid plants by hybridising and not accounting for plants used in this manner which contain odd floating chromosomes, sometimes referred to as univalents, which may cause malformation.

A cymbidium in this category is illustrated on page 79. It is one of only two cymbidiums ever awarded a First Class Certificate in Australia, the product of the immediate prewar hybridising pattern in England. Named *Cymbidium* Girrahween 'Enid', it came from Sander in about 1939. Its parents were nominated as the species *Cymbidium lowianum* with pollen parent *Cymbidium* Flamenco, an unreal pedigree, which was taken at face value.

Cymbidium Flamenco was an inbred grex using the tetraploid *Cymbidium* Alexanderi 'Westonbirt' but going back initially to *Cymbidium insigne,* together with *Cymbidiums lowianum* and *parishii.* No-one was ever very happy with the designated parentage, but it caused me to question the use of colchicine and when this really began. It is unfortunate that we do not know all the answers, but considering the wasted years during World War 2 during which so much stud and useful stock from England and Europe was dispersed, it is not surprising that these things are asked.

There were several clones of *Cymbidium* Girrahween, all good types and with at least one other which closely resembled *Cymbidium* Girrahween 'Enid'. In one of these cultivars I could see a close resemblance to a hybrid which had no connection with the grex and it was this which left me wondering.

The best clone was unbeatable for some years and retailed at a high figure for small propagations. It was one of the strongest growing plants I have ever known and fortunes have been built on this cymbidium alone. I remember seeing one plant growing in a half-cask which had eighteen spikes of flower when plants of the clone were finally allowed to become large. It needed two men to carry it into a show. Although this prime clone was used interminably in cross-pollinations, not one worthwhile cymbidium was bred from it. This signalled to me that there was an end to the road, something which I had experienced great difficulty in passing on to would-be cross-pollinators in the exhilarating years of the second wave, with everyone trying so assiduously to use triploid cymbidiums as parents. Very few ever succeeded.

There was no meristem propagation in those years and the advent of this system brought about the ultimate propagation of the genus and its spread into general horticulture in temperate climates. It is no longer an exclusive plant. Anyone may purchase it in flower and perhaps simply discard it or put it out in the garden in the shade of a tree — frequently in the hope that it may flower again.

In closing this survey of the cymbidium progenitors which provided the background for cross-pollinations of the twenty-first century, it is obvious that only a small number have been taken into account. Influences such as that of *Cymbidium* Pearl (*Cym.* Alexanderi × *Cym. grandiflorum*) have been disregarded, although they played an important role, but a very large book indeed would be needed to cover those grexes. We have a series of books which do that very well, but it is up to interested readers to put the jigsaw together.

Cypripediloideae

Cypripediums

It is difficult to separate the three genera in this group, although as far as hybridising is concerned there is little progress with true cypripediums. This is not a matter for concern, but the fact that the species has disappeared from the land surfaces it once covered freely is.

This disappearance is common to nearly all genera of orchids and the reasons are plain to see. Most destruction is caused by two groups, one human, the other animal.

I do remember one instance in an area I once regarded as 'my patch'. It contained about twenty-six species from nine genera of terrestrial orchids. Having spent some twenty-two years looking at this part of southern coastal Victoria and photographing the orchid flora, I once found a patch of pterostylis covering some square metres, the plants all in bud, some in flower. Giving it three more days to be fully open, I returned to the patch and found the plants intact but bereft of flowers. The imported rabbit had enjoyed its breakfast, lunch and dinner.

Grazing animals of all types have been the principal destroyers of more native species than is usually thought. This is perhaps one of the reasons why the true cypripedium has become a rarity in the northern hemisphere, where the genus had widespread distribution. *Cypripedium calceolus*, once a common wildflower in Britain, has been reduced to very small known surviving clumps, the remnants of human and animal destruction.

Grazing animals have destroyed many of the genus cypripedium in North America, too, where the population of a number of species is still regarded thoughtfully by appreciative people, not always orchid growers. There is interest in hybridising them, as there is esoteric Australian interest in endemic species, some of it aimed at conservation and attempted reintroduction in suitable areas. I have always regarded the latter as the tertiary stage in their destruction, unless there is a stringent conservation programme for the rest of the area in which they are replanted.

The appropriately named *Cypripedium* Genesis (*Cyp. reginae* × *Cyp. pubescens*) was processed through the Royal Horticultural Society system in 1987, thus beginning what may become another group in the already massive records of orchid cross-pollinations.

North America has the greatest number of listed species in the genus and although much of the land surface has been degraded, I have photographs of some species in situ. These were sent to me in exchange for photographs of some Australian species. Unless ecological considerations are imposed over the need to develop, the future of any species in both countries is in doubt.

Paphiopedilums

Although much of the history of this genus was presented in the first book of this series, Growing Orchids Book One, *Cymbidiums and Slippers*, further information about the species has been added to the list since that book was written.

Several decisive species not previously known, or imperfectly known, in the years of introduction to orchid collections in Britain and Europe have recently been added to the catalogue. These are either additional to known species or completely new taxa, principally from southern China. The affiliation of some of these species is obvious once you have seen *Paphiopedilum delenatii*. Other taxa are closely allied to known paphiopedilums with a reasonably long history going back to the late years of the nineteenth century.

Some new taxa exist under a cloud of suspicion that they are contrived 'new finds' with suspect origins. But the more highly regarded species such as *Paphiopedilum armeniacum* have a very bright future in cross-pollinations which will certainly reach fulfilment soon.

The first registered hybrid, as distinct from other supposed earlier cross-pollinations which are not recognised, is *Paphiopedilum* Harrisianum, with the date 1869, raised in the Veitch nursery from the cross-pollination of *Paphiopedilum barbatum* × *Paphiopedilum villosum*. The record of this hybrid in Rolfe and Hurst's *The Orchid Stud Book* is different from that in *Sander's List of Orchid Hybrids* because it named *Paphiopedilum barbatum* as the pollen parent. The grex was named by Professor Reichenbach, although Veitch is the registrant.

To those growers looking at Ball's variety (which is illustrated in Growing Orchids Book One, *Cymbidiums and Slippers*, or other later cultivars it may come as something of a shock to realise that *Paphiopedilum* Harrisianum, in the form it was named, was infinitely poorer than later cross-pollinations. One of these poor varieties was among the first Slippers in my collection, bought from an older grower and said to be a propagation from a very old plant which his father had grown in England in the late nineteenth century.

In the caption for a photograph of *Paphiopedilum* Crossianum which appeared in Growing Orchids Book One, *Cymbidiums and Slippers*, I stated that this was the second-named cross-pollination when it was really the third. The second hybrid, raised by Veitch from a cross-pollination by James Dominy of *Paphiopedilum fairrieanum* with the pollen of *Paphiopedilum barbatum*, was named *Paphiopedilum* Vexillarium in 1870.

From that point a steady though slow stream of hybrids followed from various sources, all raised principally by sowing seed on the surface of material used for potting plants of the genus. Sphagnum moss with green and growing heads was found to work well when included in potting mixes. Other methods of raising the seed were tried, but it was usually found that pots containing the genus were the best sowing beds. This slow process gave poor results in many instances and sometimes only single plants reached mature flowering size.

The first American-bred hybrid flowered in 1890. It was raised by J. Manda from *Paphiopedilum superbiens* and *Paphiopedilum concolor*. It was named *P.* Arnoldianum, and was said to have bloomed in less than two years. But in general most of the long list of cross-pollinations before the 1950s came from British and European hybridists.

There is little indication of a pattern in the following years. Rather the process of hybridising appeared to be one of haphazard cross-pollinations. The cross-pollinators do not seem to have aimed at breeding recognisable forms, as did the breeders of the mid-twentieth century, who aimed for a completely rounded outline.

Typical of the species used for cross-pollination in the late nineteenth century were *Paphiopedilum villosum* and *P. spicerianum*, which were brought into cultivation in about the middle of that century. The combination of the two produced *Paphiopedilum* Lathamianum, a grex with nineteen synonyms, later sorted through for priority and published in the *Journal of the Royal Horticultural Society* in 1888.

It may appear strange that *Paphiopedilum venustum*, the first of the genus brought into cultivation in Britain, and another early arrival in *Paphiopedilum insigne*, which arrived in about 1820, were apparently not cross-pollinated until later in the nineteenth century to produce *Paphiopedilum* Crossianum in 1873.

If one of these hybrids had to be chosen as the progenitor of the rounded outline so eagerly sought in the twentieth century it would probably be *Paphiopedilum* Leeanum, named in 1884, bred from *Paphiopedilum insigne* and *P. spicerianum*. This grex also had a long list of synonyms and a variation in outline and colour, which resulted in several cultivars being selected and given varietal status.

Competition to obtain propagations of desirable cultivars with rounded outlines was intense and they realised high prices. Among bidders for them was William Thompson, in whose collection at Stone, Staffordshire, were several which would be eagerly sought as primary hybrids by growers in the 1990s. It is doubtful if the source material responsible for their generation has survived. *Paphiopedilum* Leeanum 'Superbum' from the Veitch nursery, was considered the equal of any other but, strangely, there were apparently several forms of this cultivar.

The list of hybridists from Britain and Europe who also raised this grex and conferred different names on the product of the union is indicative of the confusion of this period. This was a common problem until society information and registration procedures were consolidated.

It should be remembered that the genus became known as paphiopedilum only in 1892, when Pfitzer's separation from the cypripediums and phragmipediums was officially recognised. Until then the three were variously named by different 'authorities'. Pfitzer began work on the three genera some ten years earlier.

The development of hybridism in this genus, which was one of the earliest introduced into artificial cultivation in Britain and Europe, was slow because of the difficult pollination techniques they needed and also because the ploidy of the genus is variable from species to species.

By the end of the nineteenth century thirty-seven paphiopedilum species had been used in cross-pollinations. Although the genus had been collected and artificially cultivated for most of that century, interest in hybridising had developed only in the years following Dominy's successful pollinations and sowing of seed. Knowing as we do the fine nature and apparent low vitality of orchid seed, it is not surprising that the growers were never very successful using techniques similar to those for garden seeds. Species which had produced hybrids which flowered and were named before the twentieth century were: *Paphiopedilums appletonianum, argus, barbatum, bellatulum, boxallii, bullenianum, callosum, chamberlainianum, charlesworthii, ciliolare, concolor, curtisii, dayanum, druryi, exul, fairrieanum, godefroyae, haynaldianum, hirsutissimum, hookerae, insigne, javanicum, lawrenceanum, mastersianum, niveum, philippinense, praestans* (*gladuliferum*), *purpuratum, rothschildianum, sanderianum, spicerianum, stonei, superbiens, tonsum, venustum, villosum.*

These species were woven in a fabric of cross-pollinations which formed the basis of the hybrid paphiopedilums grown in the late twentieth century. By the 1890s the genus had reached third generation in some grexes. Hybridists began to breed from these to achieve certain ideals. What they sought was never in the natural product at all. Even in the species *Paphiopedilum charlesworthii*, which had a rounder dorsal sepal than any previously discovered species, the petals were in every instance comparatively diminutive and other sources had to be found to develop this feature of the flower. It proved to be another series of paphiopedilums with short stems, typified by *Paphiopedilum bellatulum*, which belonged to the section known as brachypetalum, that put the final polish on the product by enhancing both dorsal and petal shape. However, it should not be thought that significant developments had not been obtained before the use of that series. Hybrids such as *Paphiopedilum* Buchanianum (*P. druryi* × *P. spicerianum*), named in 1887, gave indications of the form that motivated cross-pollinators throughout the history of the genus in cultivation.

This Slipper was not alone, nor were those parents solely responsible for the sought-after shape of the flowers. But perhaps the species *Paphiopedilum spicerianum* was the most important part of the breeding programme of nearly all the predecessors of twentieth-century Slippers. Closely following this species was *Paphiopedilum villosum*. This grex should be taken to include the species *Paphiopedilum boxallii* because they have at times been regarded as separate and at others as a single species with varietal differences. It is difficult to form an opinion on this, but there is no more difference between them than occurs between plants regarded as varieties of a species within other genera.

From the large number of pollinations and named grexes of the last half of the nineteenth century came two different forms or shapes. First was an approximation of what is known as the rounded outline and second the 'callosum' type, known from the drooping petals which are common to such species as *P. callosum*.

It took almost 100 years before cross-pollinators were satisfied that they had reached the ultimate rounded outline. This was achieved occasionally in earlier hybrids from the 1930s by manipulation of the floral parts. This involved flattening the dorsal or upper sepal by various contrivances and packing the petals with pads against the sides of the labellum or pouch with wads of cotton wool or tissues to force them back and curve the profile. Although this treatment of the flowers was not countenanced by exhibition rules, people still persisted with it. The change in the conformation of the flowers when left without this support usually showed what needed to be done to get the desired shape without manipulation.

The ventral or lower sepal also had to be rounded out to match the dorsal. This was accomplished by morphologically breeding it in, at times taking more than one generation.

Flowers which did not need this treatment to reform the bowl shape to a slightly rounded or almost flat profile were the last-stage hybrids reached towards the end of the twentieth century, but many progenitors were abandoned or discarded before this stage. This discarding process began at about the same period as the first stage in the discarding of cymbidiums, with the result that most hybrids of the nineteenth-century type almost totally disappeared.

In the strange way of human endeavour, at the end of the twentieth century species used in cross-pollinations and the hybrids themselves were being recultivated and repollinated to produce some of those earlier paphiopedilums. Some 'callosum'-type flowers have always been wanted in cultivation, either the species themselves or their hybrid forms.

One of the most attractive of these was the hybrid *Paphiopedilum* Maudiae. The grex contained one of the first so-called albino paphiopedilums, bred from the green-white forms of *P. lawrenceanum* and *P. callosum*. Originally this clone was awarded a First Class Certificate by the Royal Horticultural Society in the year 1900, but this was withdrawn when it was found that the flower could not be given to an artist for drawing. It had been pollinated and the owner was unwilling to risk it. The hybrid was raised by Charlesworths and one of the grex which had been exhibited in the previous year at Manchester, England, also gained a First Class Certificate. The form known as *P.* Maudie 'Coloratum', from the normal forms of the two species, was not raised until after the albino form was well known.

While the species *Paphiopedilum fairrieanum* was not a desired addition to collections in the first discarding phase or just after it, the grex added colour for those seeking to add red to paphiopedilums, though it tended to lose some of its distinctive shape in the process. Colour preferences vary from grower to grower but, as in cymbidiums, red flowers command attention.

Some conception of the extent of cross-pollination by the 1970s may be gauged from a pedigree extracted from *Sander's List of Orchid Hybrids*. The grex was *Paphiopedilum* Copperware, raised and named by Ratcliffe in 1968. An outsized sheet was needed to contain the 388 entries of detailed cross-pollinations and this did not list all the final species parents, which were duplicated in some instances. This is in no way extraordinary for plants with fourteen and more lines of progenitors.

A feature of such an analysis was its incompleteness because of the unknown parentage of some less used Slippers, such as *P.* Mrs W. Mostyn and *P.* Florence Spencer, which appeared several times through the pedigree. Such omissions are uncommon in later lists and unknown in twentieth-century registration procedures. In earlier cross-pollinations it was common for characteristics to be noted in primary hybrids by those who understood the morphology of the genus. Observant specialists had the ability to nominate the most likely parents and were sometimes proved right by subsequent pollinations. Secondary hybrids are more difficult. For an illustration of the mixing of floral characteristics the reproduction of *Paphiopedilum* Brunneanum on page 86 indicates the way certain characteristics never become obliterated while others become confused.

Confusion is frequently caused by failure to register seed or pollen parents before they are used to further breed a line. Most growers of any genus have at some time had plants in their collections with as many as three or more names in combination, sometimes in parentheses and sometimes not.

In detailing the pedigree or breeding table of *Paphiopedilum* Copperware, an outstanding feature was that the grex *P.* Leeanum was used over and over again. Its parents were *Paphiopedilums insigne* and *spicerianum*, raised and named by Veitch in 1884.

Although other species, such as *P. villosum*, were prominent, the dorsal sepal of *P. spicerianum* appeared to be the one which formed the basis of the roundness bred into hybrids by the end of the nineteenth century.

Once the desired flower shape had been decided, every effort was made to select parents which gave a rounded dorsal, rounded petals and a slightly concave, rather than bowl-shaped, profile. The formula was frequently subverted by nature, but when the right cross-pollinations were used even the short stems of progeny from *Paphiopedilum bellatulum*, for example, proved capable of producing longer stems in as short a time as one generation.

Two unfavourable features in the genus as a whole remain: the rather slow results from pollination to flowering and a complete inability to propagate by cloning. Despite this and the alternate waxing and waning of interest in all genera, Slippers have never suffered the decline which affected so many genera since involvement in hybridising. The history of the cypripedium — and under its later title paphiopedilum — has been as constant as that of 'cattleyas' or cymbidiums.

As in most genera cultivated for the various reasons which inspire growers, paphiopedilums have been looked at with critical eyes almost since they were introduced. This criticism developed into the ultimate appreciation rites of the late twentieth century, in which a group of judges surveys growers' entries to decide which is the 'best Slipper in the world'.

Different countries have worked out their own systems for judging these entries and it was not until the advent of world orchid conferences that there began to be any alignment. American growers were mainly responsible for organising these conferences and at the time of writing there had been thirteen seminars and exhibitions in different host countries. Appreciation of symmetry in the flowers and consensus in the management of judging the flowers leaves each country's growers still with some individuality.

Given that no original species carried the rounded form, what generated the desire in the first place — was it original development of the hybrid flowers themselves, some model forecast in the breeding programmes or human dissatisfaction with nature?

Searching through the records for leads, it appears that a few hybridists knew what they were looking for from the beginning and seemed to know how to go about generating it. The records in such publications as the continuing *Orchid Review* give an enormous amount of information, but unfortunately most of the tangible evidence has long since been scrapped. It was not so simple as it is now to present colour photographs of flowers and extant illustrations of old flowers were kept almost accidentally. Several are included here.

Each generation of cultivators thought, no doubt, that they had the ultimate in flower conformation and colour. I remember this so well from my own experience, which showed me that it was not so much the flower in the final judgement, but the ability of a grower to reproduce the full potential of the grex, also to realise that others may grow the plants better. Artificial manipulation added an imponderable.

It should be remembered, when comparing the Slippers from the mid-twentieth century with those of the 90s, that the development of a pattern brought new flowers and new ideas — particularly new ideas. If some of the older flowers are looked at with clear and unprejudiced vision it is quite apparent that what we consider outdated could be part of a never-ending and returning parade. We inevitably run out of space to accommodate this parade and thoughtful cultivators do not necessarily rush into discarding the surplus; instead they resist the temptation to buy what they cannot accommodate.

At the same point at which growers of nineteenth-century flowers reached the discarding stage, others near the end of the twentieth century find themselves taking their places in the queue. Perhaps it is the never-ending search for new colour, or perhaps only a resumption of the cycle of discarding and renewal.

It might be thought that figures are inapplicable to such things as orchids, but in this instance they tell us what is happening. Remember the list of species used in the last half of the nineteenth century, quoted on pages 13–15. Admittedly they were slowly introduced over that period and absorbed into a hybridising programme in its infancy. But in the middle of the twentieth century, only twelve of those species were used in cross-pollinations. Compare this with thirty-three species in one year, 1988, and it is easy to see the trend towards re-creation of outmoded types of flower outlined in previous pages.

While the cross-pollinations of those middle years probably reflected the ideas of the 1930–1950 period and perhaps even earlier, they indicated by their nature what was in the minds of hybridisers. Between 1946 and 1960 the grexes with the greatest number of cross-pollinations to their credit in *Sander's List of Orchid Hybrids* were *Paphiopedilum* Balaclava, a red Slipper with fifty-nine registrations and *Paphiopedilum* Diana Broughton, a green-gold with fifty-two.

Those grexes were used to produce shape and colour and numerous other cross-pollinations shared these aims. This pattern was a follow-on from haphazard cross-pollinating in the early years of the twentieth century, when hybrids or whatever species were available were used. The result of such a random system was diversity but there was little progress toward single-minded objectives.

American breeders became more involved in cross-pollination in the middle and late twentieth century than they had been at its beginning. One of the earliest hybrids was *Paphiopedilum* Amesiae (*P. fairrieanum* × *P. tonsum*), raised and registered by the American, Mrs F. L. Ames. It was first noted in *American Gardening* in 1895 as *Cypripedium* Mrs F. L. Ames. Very few other hybrids from American sources followed and in the first twenty-five years of this century hybridising was principally British and European.

In quoting the number of species used in the late twentieth century, it is notable that the only recent additions to the species tally were *Paphiopedilum armeniacum* and associated Asian Slippers. It is intriguing to speculate on cross-pollinations based on a number of other newly discovered species paphiopedilums. Most have been illustrated in the Growing Orchids series and are yet to enter breeding programmes for second or third generation plants. They display morphological differences from the flowers of so many initial species used to produce the well-formed and coloured blooms customarily noted toward the end of the twentieth century.

Will they introduce to growers a completely different set of hybrids to replace those they have been viewing in the late twentieth century? Will competitive Slipper growers need to be provided with an adjusted set of standards to embrace hybrids with a design based on different labellums and flower shapes brought into the lists?

The answer to both questions, supported by the reputation of a fickle past, is that 'fashion' will prevail and that hybridists will once more reproduce the passing parade. The fate of the discards is too well known to repeat, but who will remember them?

Phragmipediums

This third species in the cypripediloideae is cultivated far less than paphiopedilums and has little hybridising in its history. The reasons are easy to define as the flowers had no relevance in commercial orchid growing and it was perhaps natural that they did not attract growers. Although cultivation techniques for paphiopedilums and phragmipediums are similar and they are morphologically alike, it always appears that phragmipediums need slightly higher temperatures than paphiopedilums and that they are more resentful of disturbance or division.

In the history of orchids generally the phragmipediums were among early additions to collections and figured prominently in cross-pollinations performed by the innovator Seden, while Dominy's list contains only one cross-pollination in the genus and this appears under the generic title 'cypripedium'.

Seden's hybrids were also initially recorded under this title, although in later lists and registrations the plants were variously known as selenipediums, cypripediums or phragmopedilums. The author Lewis Castle did not commit himself in his book and recorded only the genus cypripedium. The separation of the three genera was not thoroughly understood until the advent of Pfitzer's distinctions in 1892, when he named the three as cypripedium, paphiopedilum and phragmopedilum.

The genus phragmipedium (Rolfe), as it is now known, is purely Central and South American, similar in many ways to the Asiatic genus paphiopedilum. Like other orchids phragmipediums come into and fall from favour in a manner which defies description. In an exhibition in Australia in the 1980s, for example, a plant of *Phragmipedium schlimii*, a Colombian orchid, was shown by a grower who had cultivated it for many years and at last encouraged it to produce a few flowers. It is pale pink and attractive and there was an immediate demand for it from other growers, undeterred by the time it had taken to grow and flower.

Hybrids within the genus were produced in the last half of the nineteenth century. In that period such an innovation was considered remarkable because of rarity and its unusual flowers. Among the earliest of these hybrids was *Phragmipedium* Sedenii, from the cross-pollination of *P. longifolium* and *P. schlimii*, a Veitch registration of one of Mr Seden's novelties.

A number of other hybrids from the twelve or more species cultivated in that period followed slowly and it would seem apparent from the paucity of follow-on hybrids that there were pollination difficulties. There are inequalities in the chromosome counts of various Slipper species and this is frequently a barrier to inter-generic hybridism, despite 'look-alike' qualifications.

Cross-pollinations with the other genera were tried, but cross-pollination with the cypripedium once common to British woodlands proved intractable, and nothing ever came of those attempts. Pollination between paphiopedilums and phragmipediums was more successful and plants were brought to maturity by Veitch, but they refused to flower. In the continuing history attempts have probably been frequent and persistence rewarded in few instances, although seed capsules have occasionally been set.

The attempt to cross-pollinate the genera paphiopedilum and phragmipedium offers hybridists the greatest challenge in orchid growing. In more than 100 years only two successful cross-pollinations have been registered with the Royal Horticultural Society in *Sander's List of Orchid Hybrids*. This may not seem unusual because there have been so many other ineffectual attempts at what is, in effect, miscegenation, but with three genera so much alike, more could have been expected.

The first hybrid was registered as *Selenocypripedium* Malhouitri in 1912, from the cross-pollination of *Paphiopedilum* Harrisianum and *Phragmipedium schlimii*. The registrant was listed simply as Boullet. *The Orchid Review* of March 1913, commented:

> At a meeting of the Société nationale d'horticulture de France held on November 28th last a curious hybrid, said to have been obtained by crossing *Cypripedium* Harrisianum with the pollen of *Selenipedium schlimii*, was exhibited by M. Eugene Boullet, of Corbie, under the name *Selenocypripedium* Malhouitri . . . Although not of much interest from a decorative standpoint, it is said to quite dispel a general idea that the two genera cannot be crossed together. Only two seedlings were obtained, one of which died when very young, so that the plant is at present unique. We should be very much obliged for a flower of this interesting plant, for previous crosses between the two genera which we have seen did not show any hybrid character. M. Boullet is said to have a capsule nearly mature on *C.* Leeanum crossed by *S.* schlimii.

Looking carefully at the full quote, which described the flower in detail, it is obvious that it was intermediate between the two parents. It is also obvious that other so-called cross-pollinations had been flowered but did not show a pattern of hybridism. The 'we' in the closing remarks of the quote most probably referred to Rolfe, the editor.

The second hybrid was registered in 1983 as *Phragmipaphium* Hanes' Magic from the cross-pollination of *Paphiopedilum stonei* and *Phragmipedium albopurpureum*. The reason for changing the already adopted intergeneric title 'selenocypripedium' in volume one of *Sander's List of Hybrid Orchids* has no logical reason other than the change in name for the cypripedium. The registrant was Hanes Orchids, of California, USA, an American nursery which became prominent in the breeding of paphiopedilums in the later half of the twentieth century.

Judging by the emergence of this second hybrid it is possible that further work on the genera could be rewarding so long as there was a large market, such as that of the USA, to make it worthwhile. There would also need to be a change in fashion. But for the hybridist who can breed the red *Phragmipedium besseae* into the complex a new door will open.

Dendrobiums

The dendrobium genus is among the largest in the world, but its true extent is unknown. We need to realise that hybridism is a continuing thing in nearly all orchids with sizeable populations, such as dendrobium. The plants are Indo-Asian, as are cymbidiums, and where one genus occurs we may also find the other. The total range is from northern India to some of the western Pacific islands, such as Fiji, New Zealand and the Solomons.

The morphological variation over this large area is similar to the differences between cattleyas and lycastes. Growers knowing the difference between *Dendrobium linguiforme* and *Dendrobium nobile* would realise this. For this reason artificial hybridism or cross-pollination must be treated as belonging to two sections — the soft-cane and the hard-cane species which, although they are usually incompatible, may mix. This is rough categorising of a principally epiphytic genus. Some species fall between the two, but it is not simple to classify and describe these imponderables.

The country with the most undescribed and unclassified dendrobiums in the late twentieth century is probably New Guinea, taken as a whole. Soft-cane, hard-cane and the hirsute 'in-between' dendrobiums all grow there.

Morphology, or plant design, is most diverse. Two Australian species, *Dendrobium speciosum* and *Dendrobium toressae,* can be considered as examples. The first is principally tall growing and stout, whilst the other is diminutive, with pseudo-bulbs resembling tiny leaves strung like beads on a rhizome with bulbophyllum-like morphology.

Cross-pollinations within the two sections, hard-cane and soft-cane, are sometimes curious, but very few represent misdirected effort. Considering the size of the genus, cross-pollinations have been relatively few and mostly made with the purpose of broadening commercial use. To those who do not cultivate dendrobiums it may sometimes seem that nothing has been gained, for example, with the *Dendrobium nobile* complex. In most instances, however, a breakthrough has brought new growers into the section after a lull of many years. They are attracted by the flowers of an apparently new vision.

In most instances new and brilliant soft-cane hybrids of Asian origin, such as the Yamamoto hybrids, were bred from stock which had been 'put on the shelf' many years previously. They brought thousands of cultivators who had a lot to learn in that section. Some succeeded, but for each one who did so there were others who could not supply the environment needed for plants to flower. In fact, very few reached ultimate success.

Having begun to look at the soft-cane dendrobiums, we will proceed through the various stages which led to the so-called 'Yamamoto' hybrid flowers. That description may not be strictly correct, but it does refer to the flowers of a 'new breed' and most readers will understand the difference between the flowers by looking at the illustrations in this and the other books in the Growing Orchids series.

Soft-Cane Dendrobiums

Most genera have an initiating species, such as *Cattleya loddigesii* for cattleyas or *Paphiopedilum spicerianum* for that genus. *Dendrobium nobile* could be regarded as the principal progenitor or initiator of the soft-cane hybrid line and most dendrobium growers have had at least some success at cultivating it and having it flower. The wise ones still have this simple and easy-to-cultivate plant. Unfortunately most beginners start by buying the plant first and then trying to find out about the correct environment for it as they go along. The environment requirements have been frequently noted elsewhere and in this book we will concentrate on the results.

In nominating *Dendrobium nobile* as the starting point I have taken a lead from John Lindley's book *Sertum Orchidaceum*, published in 1838, in which a beautiful lithograph of this plant appears. A plant sent to Loddiges' nursery in Britain from Macao flowered in 1837, but before that a drawing of the species in flower was brought to England. It was said to have been drawn from a plant found in southern China. Knowing the richness of the cool-growing orchid flora which originated in that general area, including parts of South-East Asia and Burma, it is not surprising that further plants were sought and obtained.

It was some time, however, before much attention was given to pollinating the flowers, which is understandable when we consider the rarity and newness of the orchid. By the time the first cross-pollinations were made, other soft-cane dendrobiums were flooding into the British and European glasshouses. Most of these, however, came from Burma or the Shan States not very far west of the original source. Some were of the same or similar species and it is possible that the total area over which the plants grew was not known.

While *Dendrobium* Ainsworthii receives credit as the first hybrid generated from this soft-cane line, apparently it is not entitled to the honour, although it flowered and was named in 1874. James Dominy certainly cross-pollinated and raised seedlings from *Dendrobium nobile*, fertilised with the pollen of *Dendrobium linawianum*, some years before that, but they were not named or registered until 1878 as *Dendrobium* Dominyanum. The pollen parent originated mostly in China and Japan and occasionally occurred in other cool localities. It is now known as *Dendrobium moniliforme*.

Dendrobium heterocarpum, syn. *D. aureum*, was the seed-carrying parent of the hybrid *Dendrobium* Ainsworthii, a progenitor for most soft-cane hybrids to the end of the twentieth century. *Dendrobium heterocarpum* is widespread, but it was mainly the bright gold Burmese variety which added colour to hybrids.

This species also gave rise to *Dendrobium signatum*, the mystery of the hybrid register. No more is known about this orchid at the end of the twentieth century than when speculation about its origin began. Sir Jeremiah Colman perhaps knew more than he revealed, but to our great misfortune he never disclosed much. Further information is given in Growing Orchids Book Three, *Vandas, Dendrobiums and Others* in the section on dendrobiums.

The future of hybridism based on *Dendrobium nobile* was changed with the introduction of the variety *nobilius*, although this was not the only fine variety which followed the introduction of the taxon. Several others were known in the late nineteenth century, among them *Dendrobium nobile* var. *sanderianum*, illustrated in *Reichenbachia*. As with many genera, there appeared to be an endless number of forms of *Dendrobium nobile*. Most of these known and used varieties have disappeared, but the variety *nobilius* is still cultivated. The essential differences between this and others were more intense colour and slightly larger size. The pure white form, known as variety *virginale*, also appeared in the late nineteenth century.

Most forms of *Dendrobium nobile* had a serious drawback for cross-pollination — the flowers tended to fall away from the main stems into an almost horizontal position in which it was very difficult to see into them without lifting the pedicel of individual blooms. This fault is uncommon in cross-pollinations of selected clones or with other species which do not carry this habit and present flowers with better aspects.

A feature of *Dendrobium nobile* flowers is the enduring patterns of the blooms, almost regardless of other species used in cross-pollinations. About the only major change in configuration is a broadening of flower segments to the point where the profile, including the lobes of the labellum, is almost flat.

Other species which had an effect on this transformation included *Dendrobiums crystallinum, primulinum, pierardii, wardianum, findlayanum, falconeri, lituiflorum, crassinode, hildebrandii, bensoniae, parishii, crepidatum* and one or two others used only infrequently. If the colour combinations of some could have been preselected, the rainbow effect of the late twentieth century may have been accentuated.

Most initial development to secondary hybrid generation occurred in the last half of the nineteenth century and, from then on, more emphasis was placed on altering the shape of flowers to conform to the 'circular' outline which seemed fixed in hybridists' eyes as an ultimate goal.

In that period several alliances with hard-cane and intermediate type dendrobiums occurred, but these were relatively unimportant except in those involving species such as *Dendrobium hildebrandii*. This dendrobium was brought into cultivation from Burma and, with *Dendrobiums heterocarpum* and *bensoniae*, had effective use in breeding yellow to gold tones into hybrids. Further information on their use is given in Growing Orchids Book Three, *Vandas, Dendrobiums and Others*.

Subsequent introduction of different species had little effect on the development of soft-cane dendrobiums and there was little or no reworking of older cross-pollinations to produce higher quality clones than the originals. Perhaps this was because there was more than enough material to further the aims of hybridists in the select clones of the original.

'Fashion' dictated the fate of soft-cane cross-pollinations and cultivation of progeny in the same way as for the other cultivated genera. While the nucleus survived to bring the genus back into favour at any time, many growers deserted the soft-cane family for other genera or newer hybrids.

Although it is difficult to assess progress in the first half of the twentieth century realistically, it seems fair to say that progress with soft-cane dendrobiums consisted of slow, inbred steps towards an ill-defined goal.

The species *Dendrobium nobile* var. *nobilius* still formed the basic ingredient of programmes carried out by a few enthusiasts, but the total number of soft-cane and hard-cane dendrobiums registered between 1874 and 1946 was under 400. Compared with other genera such as paphiopedilums (a fair comparison since they are both monotypic genera), this clearly illustrates how small a proportion of the total number of cultivated orchids they represented.

It may be appropriate to recall that immediately after World War 2 thousands of plants of the hybrid *Dendrobium nobile* var. *cookson's* × *Dendrobium regium* were exported to Australia, principally in community-pot lots. This gave a 'new-look' hybrid with a light-coloured central lobe on the labellum, but it is now difficult to find many of those plants in Australia.

This was the inception, however, of a gradual increase in cultivation of the genus, soft and hard cane alike. In considering soft-cane dendrobiums it should be realised that several grexes, some far from new, began to impress growers from the 1950s to the 1990s. Many of these had a light interior on the labellum. Others, such as *Dendrobium* Montrose, were yellow-gold types, mostly of only moderate shape, bred from *D. heterocarpum* and *D. bensoniae* hybrid lines.

Much of this interest would never have occurred had it not been for increases in the number of orchid growers and the establishment of annual exhibitions after 1950 by societies and orchid-growing communities almost worldwide. These exhibitions brought better cultivation techniques because of competition between growers and this, in turn, brought more growers into the pastime.

At the same time adventurous enthusiasts began experimenting with soft-cane hybrids and found that open-air cultivation suited them. It is salutary to realise, however, that in some instances this occurred where plants were shifted outdoors to take their chances in less than ideal environments. They succeeded beyond expectation, but flowering was never as good as when they were grown in better conditions where there were no violent fluctuations in temperature.

In the last quarter of the twentieth century the so-called Yamamoto hybrids, principally based on soft-cane dendrobiums developed in British and European nurseries many years previously, burst onto the scene. The originators apparently found them hard to place in countries dependent on glasshouse cultivation needing several months of heating back-up.

Salesmanship and commercial marketing succeeded in popularising the Yamamotos and they soon created a demand amongst those fortunate enough to see well-flowered plants.

As in all orchid growing, it is necessary to live in or develop suitable environments in order to replicate the beautiful displays these soft-cane dendrobiums can produce. You need space, free air flow, and moderate warmth, preferably combined in a natural climate.

In following the pedigrees of nearly all soft-cane hybrids we find entries for the same species occurring over and over again, whatever the order in which they were introduced.

Dendrobium Malones (*D.* Akatuki × *D.* Glorious Rainbow), a particularly fine grex from the Yamamoto group, exhibits most characteristics sought by hybridists and fanciers alike. Under bloodline analysis it shows a continual use of derivatives from *Dendrobium nobile*, which appears sixteen times through the pedigree, *Dendrobium aureum* (so-called in earlier lists but known correctly as *D. heterocarpum*) nine times, *D. findlayanum* six times, *D. signatum* twice, *D. hildebrandii* once, *D. crassinode* once and *D. linawianum* (*D. moniliforme*) twice.

In this pedigree, as in many hybrid records, there is a doubtful entry. *Dendrobium regale* is listed as a species in the 1961–1970 issue of *Sander's List of Orchid Hybrids*. It was once in my collection as a hybrid, but always foiled my attempts to find a source. It appears only in the initial 1946 catalogue, in which it is described as a parent, with *Dendrobium* Euryalus, of *Dendrobium* Pilgrim. A possible source exists for it as an unregistered hybrid, but search through the first fifty years of *The Orchid Review* gives no record of *Dendrobium* Regale, which is most unusual.

Dendrobium Malones is only one of many hybrids in the so-called Yamamoto 'soft canes'. The colours of the whole style or breed, ranging through white and yellow-gold to deep red-purple, were brought into the catalogue from the 1900–1930 period of soft-cane cross-pollinations. *Dendrobium* Kobayashi (*D. nobile* × *D. Merlin*), named in 1968 by Jiro Yamamoto of Japan, is one of the progenitors of *Dendrobium* Malones. *Dendrobium* Kobayashi has been among the soft canes in my collection since the end of World War 2. Its attraction, as previously stated, is that the flowers are held on the pseudo-bulbs in such a way as to make it unnecessary to lift them to see into the throat of the labellum.

It is significant that, after a hybridist raised *Dendrobium* Kobayashi as seedling stock, more than twenty years elapsed before it was named, when it was used as a seed parent for registration of its scions. Although its pollinator was untraceable, it was possibly a result from the early twentieth century. The fact that it was unnamed and disused for so long indicates that this section of the genus was partly abandoned or fell out of fashion. The pseudo-bulbs of *D.* Kobayashi frequently grow to more than a metre (about four feet), a height few of the soft-cane species ever attain. It also flowers freely.

△ *Cattleya* × Hybrida (*C. granulosa* × *C. harrisoniae*), credited to James Dominy through the Veitch nursery, the first hybrid cattleya. Although noted in *Sander's List of Orchid Hybrids* as a natural hybrid, it is difficult to take from Dominy the honour of at least confirming its status.

▷ *Cattleya trianaei*, a Colombian unifoliate, probably had the greatest influence in hybridists' search for 'shape'. There is sufficient evidence in early progenitors such as *Cattleya* Edithiae to validate the claim.

△ *Cattleya dowiana* var. *aurea*. As detailed in the text, it is unfortunate that the records of the two varieties of *Cattleya dowiana* were not kept separate. (Photograph courtesy of Barbara Walker.)

△ *Cattleya* × Hardyana. Early hybridists had little fortune trying to intensify the yellow into gold in either the parents or this natural hybrid from their cross-pollinations.

◁*Cattleya warscewiczii* syn. *C. gigas*, with *Cattleya dowiana* var. *aurea*, became a strong influence in cattleya cross-pollinations through the natural hybrid *Cattleya* × Hardyana. These three Colombian cattleyas are part of the basic pattern of hybridism.

▷*Cattleya* Enid (*C. mossiae* × *C. warscewiczii*), raised by Veitch and named in 1898, is among the outstanding early cattleya hybrids which provided background to better cross-pollinators at the end of the twentieth century. Such cattleyas are timeless.

▽ *Cattleya* Paula Marabella, produced towards the end of the twentieth century, still carries the shape and colour of its far-off progenitors.

△*Cattleya trianaei* var. *alba*, together with albino forms of *Cattleyas mossiae* and *gaskelliana*, probably produced a better background in hybridising material than the coloured forms. This is borne out by an examination of breeding lines back to the originals.

▷*Cattleya* Princess Bells. A development of the albino line, mentioned in the text. While there are equally finely shaped cattleyas in coloured forms, these whites almost monopolised the attention of those who grew the genus at one time.

△*Cattleya* (unnamed) (*C.* Barbara Kirsch × *C.* Chocolate Drop). Indicative of the trend in the late twentieth century towards smaller flowers, provided they carried bright colour.

▷*Cattleya* Mantinii (*C. bowringiana* × *C. dowiana*), a 'cluster' type bred from smaller species, raised in 1894. It represents the blending of unifoliate and bifoliate species and set a pattern for exhibition-type, mid-sized flowers with greater numbers than pure unifoliate strains.

▽*Brassocattleya* Pastoral (*Cattleya* Mademoiselle Louise Pauwels × *Brassocattleya* Déesse) illustrates the full labellum development of the species *Brassavola* (*Rhyncolaelia*) *digbyana* in the 1960s.

△*Brassocattleya* Gatton Snowflake (*Bc.* Thorntonii × *Cattleya chocoensis* var. *alba*), named in 1921, representing an intergeneric series of small numbers compared with other 'cattleya' entries in *Sander's List of Orchid Hybrids*.

▽*Brassocattleya* (*C. loddigesii* × *Brassavola* [*Rhyncolaelia*] *glauca*). It is obvious why this brassavola species was not much used in hybridising compared with its more decorative relative.

△ *Brassolaeliocattleya* American Heritage (*Blc.* Fortune × *Blc.* Golden Slippers), a complex product of the 1970s. The overall colour line reaches back to the species *Cattleya dowiana* and *Laelia tenebrosa*, both strong influences for yellow to gold.

△▷ *Brassolaeliocattleya* Nacouchee × *Cattleya* Esbetts. An unregistered grex, probably from the last half of the twentieth century. The shape illustrates the overbearing influence of *Cattleya* Bob Betts and, through it, to *Cattleya* Edithiae (*Cattleya* Suzanne Hye × *Cattleya trianaei*).

▷ *Brassolaeliocattleya* Waikiki Sunset (*Blc.* Walter Abe × *Lc.* Waianae Sunset). The softness of the brassavola colouring and delicacy of the labellum are reflected in this Hawaiian hybrid of the middle years of the twentieth century.

▷*Brassolaeliocattleya* Yellow Ribbons 'Starfire' (*Lc.* Ann Folis × *Blc.* Golden Destiny). A Californian hybrid, perhaps the shape, style and colouring of things to come. (Photograph courtesy of McKinney's Nursery, Sunnybank, Queensland.)

▽*Laeliocattleya* Parysatis (*C. bowringiana* × *L. pumila*), 1893. In the beginning there were only species, now few growers have them in their collections. Possibly fewer still have primary hybrids such as this.

▷*Laeliocattleya Fedora* (*Lc.* Laguna × *C.* × Hardyana), 1931, brought an award to the author in the middle years of the twentieth century. It represented a fine early development of laeliocattleyas in the short period since inception, using only *Laelia purpurata*.

△ *Laeliocattleya* Memoria Georgina Nevins (*C.* Gene May × *Lc.* Kindee),1986. Bifoliate cattleya background was responsible for this brilliant small 'cattleya', part of a series popular in the middle to late twentieth century.

▷ *Sophrocattleya* Doris (*Cattleya dowiana* × *Sophronitis grandiflora*, commonly referred to as *S. coccinea*). There is some uncertainty, considering the changed nomenclature of the group, which sophronitis was the pollen parent of this hybrid, which was registered in 1904 and is still in cultivation.

◁ *Sophrolaelia* Psyche (*Laelia cinnabarina* × *Sophronitis coccinea*). This miniature was a product of 1902 but retains its place in hybrid ranks in the late twentieth century and has been used extensively as pollen or seed parent.

▷ *Sophrolaelia* Jinn (*Sophronitis coccinea* × *Laelia milleri*). *Sl.* Jinn appeared in 1966, one of the first hybrids generated after the discovery of *Laelia milleri* and its entry into cultivation. Sophronitis as a genus is indistinctly defined, as outlined in Growing Orchids, *Expanding Your Orchid Collection*.

▷ *Sophrolaelia* Orpetii (*Laelia pumila* × *Sophronitis coccinea*). Probably the best of the sophronitis-laelia combinations, raised and named in 1901. Remakes have been infrequent, but a repeat in the last half of the twentieth century provided colourful, cool-growing, red miniatures.

◁ *Sophrocattleya* Batemaniana (*C. intermedia* × *S. grandiflora*), 1888. Named after a man who detested orchid hybridism, this cross-pollination used *Cattleya intermedia* var. *aquinii* where the original used the ordinary form.

△ *Sophrolaelia* Marriottiana (*Laelia flava* × *Sophronitis coccinea*). A hybrid named and registered in 1896, one of the first miniature 'cattleyas'. These small hybrids came back into vogue in the last half of the twentieth century, but comparatively few outshone the intense colour of the originals.

▷ *Sophrolaeliocattleya* Falcon 'Westonbirt' (*Laeliocattleya* Aureole × *Sophronitis grandiflora*). This hybrid was probably created by H. G. Alexander when detailing work in the Holford collection in England. It was registered and named in 1917, and is about 65 millimetres. Its brilliance may be equalled by others, but not surpassed.

▷ *Cattleytonia* Rosy Jewel (*Cattleya bowringiana* × *Broughtonia sanguinea*). The appearance of the Jamaican broughtonia marked the middle of the twentieth century as a period of experiment.

▷ *Potinara* Apocalypse (*Pot.* Gordon Siu × *Sophrolaeliocattleya* Tropic Flare), 1973. It is seldom a good idea to intensify colour in an already bright line, but this success in a moderate-sized flower demonstrates the full breeding potential of sophronitis.

▷ *Potinara* Hawaiian Landmark (*Brassolaeliocattleya* Waikiki Sunset × *Sophrolaeliocattleya* Kauai Starbright), 1989. A complex hybrid in the 'cluster' form which became popular in the last half of the twentieth century.

◁ *Laelia* Latona (*L. cinnabarina* × *L. purpurata*). Raised by Veitch and named in 1892, this grex was little used throughout its history. Perhaps the unhappy mix of two distinct types, it did not carry the potential which flowed from laeliocattleya cross-pollinations.

▷ *Laelia milleri* × *Schomburgkia* (*Laelia*) *superbiens*. An unnamed cross-pollination of Australian origin. Free growing and bearing flower racemes a metre tall, it is most colourful. The flowers are eight to nine centimetres in diameter. This is a subject for cool to intermediate climates.

▷ *Laelia* Cinnabrosa (*L. cinnabarina* × *L. tenebrosa*). A brilliant combination between Brazilian species, spring flowering and preferring a moderately warm climate. The combination is perhaps a 'touchy' hybrid which some growers may find difficult but it is possible to grow it in cool to intermediate environments.

◁ *Cattleytonia* Keith Roth × *Dialaelia* Snowflake. The breeding of the pollen parent is *Diacrium* (*Caularthron*) *bicornutum* × *Laelia albida*, a grex not registered until mid-1990. Six genera bonded in a hybrid was the highest figure in that year.

△ *Brassocattleya* Makai (*Cattleya bowringiana* × *B. nodosa*). A reverse-order hybrid which followed the floriferous habit of the bifoliate cattleya. The inherent characteristics of hybrids generated from brassavolas show trends toward terete-type foliage rather than the lanceolate leaves of cattleyas.

▷ *Brassoepicattleya* (?). A cross-pollination not recognised in the list as this book was written. It is the combination of *Brassoepidendrum* (*Bepi.*) Phoenix × *Cattleya guttata*. These out-crosses from recognised patterns are not assured of a future unless they have outstanding attributes.

▷ *Brassocattleya* Star Ruby (*Brassavola nodosa* × *Cattleya* Batalinii). *Cattleya* Batalinii is a natural hybrid which was identified by Sanders in 1892. Its parents are *C. bicolor* and *C. intermedia*. This type of orchid breeding was increased in the late twentieth century.

△ *Brassocattleya* Nodata (*B. nodosa* × *C. guttata*). A hybrid which appeared in the middle of the twentieth century and has continued to attract cultivators of miscellaneous orchids since then.

▷ *Cymbidium* Coningsbyanum (*C. grandiflorum* × *C. insigne*), 1914. This combination brought *C. grandiflorum*, at 115 millimetres the largest of the species, into the breeding lines of further hybrids. *C. grandiflorum*, the subject of part of Growing Orchids Book One, *Cymbidiums and Slippers*, was the background to most of the green hybrids from the beginning of that phase of cymbidium culture.

△*Cymbidium* Pauwelsii 'Comte de Hemptinne', 1911. This is the source of colour for the Westonbirt breeding line and it has endowed the bloodline with large flowers, principally owing to genetic doubling, or 4n status, a quality shared by *Cym.* Alexanderi Westonbirt variety.

▷*Cymbidium* Swallow (*Cym.* Alexanderi × *Cym.* Pauwelsii), 1916. This photograph shows the colour and form of the 4n hybrid. There was also a 3n form bred with *Cym.* Alexanderi (4n) and *Cym.* Pauwelsii (2n — normal). Little or no cross-pollination was effective with 3n forms, but 4n clones in correct combination led to larger hybrids.

▷*Cymbidium* Ceres (*Cym. i'ansonii* × *Cym. insigne*), 1919. Most pink and red cymbidiums owe their existence to the colour line introduced through this cross-pollination. There was considerable variation in shape and size, but the two species carry colour factors which are scarcely present in other cymbidiums.

◁*Cymbidium* Alexanderi (*Cym.* Eburneo-lowianum × *Cym. insigne*) Westonbirt variety, 1928. It is not overstating the importance of this clone to give it most credit for outstanding cymbidiums at the end of the twentieth century. (Photograph courtesy of Frank Slattery.)

◁ *Cymbidium* Cyzara 'Remembrance' (*Cym*. Albanense × *Cym*. Ceres). The original red-brown of *Cymbidium* Ceres was in transition to clearer colour, although it was almost impossible to eliminate the striping of colour in petals and sepals in further cross-pollinations.

▷ *Cymbidium* Rosanna 'Pinkie' (*Cym*. Alexanderi 'Westonbirt' × *Cym*. Kittiwake). A 1927 hybrid which revolutionised the genus and became the paramount carrier of the tetraploid characteristic. (Photograph courtesy of Frank Slattery.)

▷ *Cymbidium* Panama Red (*Cym*. Sensation × *Cym*. Khyber Pass), 1978. The intention to produce rounded shape with intense colour has been somewhat thwarted by the lining of petals and sepals, an inherent characteristic of the 'red line'. This grex also flows back to *Cymbidium* Ceres and through it to the species *Cym. insigne* and *Cym. i'ansonii*. Identifiable characteristics from these are the colour, the red lining of the interior of the labellum, the red tip of the column and the intense red banding of the labellum. *Cym. tracyanum* also added red-brown to the overall markings and colour.

▷ *Cymbidium* Dorchester 'Janette' (*Cym*. Alexanderi 'Westonbirt' × *Cym*. Tityus), 1932. This awarded clone held a top position for a long period. It will probably remain the classic form and colour of the original Westonbirt line of the genus in triploid matrix.

◁ *Cymbidium* Louis Sander 'Picardy' (*Cym*. Alexanderi × *Cym*.Ceres). First raised as a hybrid in 1924 with normal 2n parents, later repeated with *Cym*. Alexanderi (4n) and *Cym*. Ceres (2n), giving a 3n or triploid hybrid, mostly sterile but vigorous. This clone gained an award for the author in 1949.

△*Cymbidium* Euterpe 'Churchill' (*Cym.* Lowio-schroderi × *Cym.* Swallow), 1936. This was the first cymbidium to gain an award in Australia when Richard Dart was given a Highly Commended Certificate for it, in New South Wales in 1944. (Photograph courtesy of Frank Slattery.)

◁*Cymbidium* Joyance 'Cinnamon' (*Cym.* Joy Sander × *Cym.* Pauwelsii), 1937. From a *Cymbidium* Pauwelsii 'Comte de Hemptinne' cross-pollination, probably a triploid with remarkable colour and carrying more than fourteen large flowers. Probably no longer cultivated.

▷*Cymbidium* Remus FCC (*Cym.* Regulus × *Cym.* Joyful), 1939?. A cymbidium about which there was much confusion and the parentage came into doubt. It had little impact on cymbidium hybridism compared with other grexes, producing a few outstanding hybrids, principally 3n, such as *Cymbidium* Burgundian. (Photograph courtesy of Frank Slattery.)

▽*Cymbidium* Girrahween 'Enid' (*Cym. lowianum* × *Cym.* Flamenco), 1941. Probably *the* remarkable cultivar in the first half of the twentieth century in Australia. It gained a First Class Certificate in 1944. (Photograph courtesy of Frank Slattery.)

◁*Cymbidium* Minuet (*Cym. insigne* × *Cym. pumilum*), 1942. A grex notable as the forerunner of a line of hybrid miniature cymbidiums. The first recorded miniature raised and named was *Cymbidium* Dingleden in 1933 from the parents *Cym.* Alexanderi × *Cym. devonianum*. It was a very dark, attractive flower, but lacked follow-up, probably because of 3n ploidy and thus sterility.

▽*Cymbidium* Lunagrad 'Elanora' (*Cym.* Miracle × *Cym.* San Miguel). The difference genetic interference makes to cymbidiums is illustrated in this combination. The smaller flower is from a natural meristem, the larger is from a colchicine-treated meristem. The identifiable characteristics from species level are the modified red banding of the labellum from *Cym. lowianum*, the speckling above the banding and on the base of the column from *Cym. grandiflorum* and the colour from both these species.

△ *Cymbidium* Showgirl (*Cym.* Sweetheart × *Cym.* Alexanderi), 1955. This miniature, probably now recognised as an intermediate in some ratings, was almost the beginning of the vogue for smaller flowers. Although in recent decades the 'intermediate' almost dominated the scene, it never took over from standard cymbidium hybrids successfully.

▽ *Cymbidium* Sweet Lime (*Cym.* Esmeralda × *Cym. madidum*), 1964. The pollen parent, an Australian, is among the smallest species of cymbidiums. Its use accented the green of *Cym.* Esmeralda but lost the trailing raceme of both parents.

△ *Cymbidium* Olive Street (*Cym.* Darjeeling × *Cym. devonianum*), 1986. Use of *Cym. devonianum* brought excellent colour to the whole flower, with intensity in the labellum. The 'spidery' outline is of no great consequence because most miniature cymbidiums are grown for their novelty.

◁*Paphiopedilum* Harrisianum (*P. barbatum* × *P. villosum*), 1869. The first Slipper hybrid, raised by Veitch after the cross-pollination by James Dominy. A common form, not the darker Ball's variety. Other references to this hybrid and the genus appear in the other books in this series. All pre-1892 Slippers were registered as cypripediums but transferred to paphiopedilum by Stein in his rearrangement in that year. (See Growing Orchids, *The Specialist Orchid Grower*.)

▽ *Paphiopedilum* Oenanthum (*P.* Harrisianum × *P. insigne*), 1876. Probably the first registration of a secondary hybrid, credited to Seden, who was no doubt inspired to use *P.* Harrisianum. Named by the Veitch nursery, it was also flowered by others and twenty-eight references to the hybrid appeared in various journals, some with names different from the registration.

△ *Paphiopedilum* Ashburtonae (*P. barbatum* × *P. insigne*), 1871. This was probably the second primary hybrid, not *P.* Crossianum, as stated in Growing Orchids Book One, *Cymbidiums and Slippers*. It was raised and named by Ashburton, an English grower, and registered in the *Journal of the Royal Horticultural Society*.

△*Paphiopedilum* Io (*P. argus* × *P. lawrenceanum*), 1886. Certainly the shortest name for a hybrid, it was raised and named by Cookson and registered in the *Gardener's Chronicle*. Cross-pollinations of the late twentieth century have a callosum-like similarity to it.

▷ *Paphiopedilum* Leeanum (*P. insigne* × *P. spicerianum*), 1884. One of the important early hybrids because it formed a principal background link with hybrid paphiopedilums of the late twentieth century. It was originally raised by Veitch, but a number of other cross-pollinators achieved parallel results, many with different names for it.

◁*Paphiopedilum* Alice (*P. spicerianum* × *P. stonei*), 1890. A type of hybrid referred to in the text as having attraction, perhaps as a novelty. But once placed in collections such Slippers are difficult to discard, considering the overall diversity common at the end of the twentieth century.

◁*Paphiopedilum* Calloso-barbatum, 1890. There is little need to give the parentage of this hybrid, since it is part of a common pattern in orchid collections and one which is not difficult to replace with more attractive cultivars.

◁*Paphiopedilum* Tautzianum (*P. barbatum* × *P. niveum*), 1886. This was a very early use of the brachypetalum section of the genus, most members of which have compact flowers. *P. niveum* has longer stems than others of the group and was a leader in the trend toward white hybrids.

▷*Paphiopedilum* F. C. Puddle (*P.* Actaeus × *P.* Astarte), 1932. One of the nearest approaches to pure white ever raised. Growers are constantly seeking larger hybrids with this clearness and, if the project is started correctly, breeding out colour intrusions is less difficult.

▽*Paphiopedilum* Conco-bellatulum, 1891. A hybrid between two members of the brachypetalum section. A repeat of this cross-pollination, but using albino forms of the species, should be part of the amoury of hybridisers for an attack on the pure white paphiopedilum line.

◁ *Paphiopedilum* Brunneanum (P. Leeanum × P. Oenanthum), 1892. Referred to in the text as a combination of different characteristics. This Slipper is a secondary hybrid carrying direct lines of *P. barbatum*, *P. insigne*, *P. villosum* and *P. spicerianum* and a careful analysis of the flower shows something from each of these species.

▽ *Paphiopedilum* Olivia (*P. niveum* × *P. tonsum*),1898. Despite attractiveness, only four hybrids, one of which used *P. sukhakulii*, were generated from this hybrid in almost 100 years. Considering the alliance of this attractive hybrid with that species leads one to conclude that novelty was the aim.

▷*Paphiopedilum* Redstart (*P.* Nubian × *P.* Mrs Cary Batten). This parentage is recorded in the *Orchid Review* of 1932, although there is no entry in *Sander's List of Orchid Hybrids*. Both parents have unknown progenitors, but *Paphiopedilum charlesworthii* suggests itself as a principal candidate. *Paphiopedilum* Redstart is the most notable red antecedent of all.

▷*Paphiopedilum* Battle of Egypt (*P.* Miracle × *P.* Redstart), 1942. A step on the way towards the rounded flower, still using the red *P.* Redstart 'Exbury.'

▽*Paphiopedilum* Maudiae 'Coloratum'(*P. callosum* × *P. lawrenceanum*). Usually noted as an albino or green and white Slipper of the year 1900, this coloured form in fact appeared six years later and was a notable addition to the 'callosum' type hybrids leading to the 'vinicolour' section in 1970–80.

△*Paphiopedilum* Thule (*P*. Altair × *P*. Windrush), 1936. This spotted form also has some brushing and a scarcely surpassed shape for its time.

◁ *Paphiopedilum* Upton Gem (*P*. Alcibiades × *P*. Idina), 1922. In general terms there are three patterns exhibited on the dorsal or upper part of Slipper flowers — spotted, solid colour or brushed. *P*. Upton Gem represents the spotted form. It is also an almost full development of the required shape, achieved early in the twentieth century.

△ *Paphiopedilum* Feldspar (*P.* Bordube × *P.* Gorse), 1964. Subtle tinting, pleasing overall colour and the desired shape make this clone top quality. The centre dark line was inherited from *P. spicerianum* many generations back in the pedigree.

△◁ *Paphiopedilum* Gaystone (*P.* Blondel × *P.* Tafel Rose), 1975. A good example of the brushed form, which may be any colour, preferably in a pleasing combination if mixed. The principal attribute is the circular outline and overlapping parts of the whole flower. Reference to the illustration of *P.* Orchilla gives the best example needed for solid colour.

◁ *Paphiopedilum* Orchilla 'Chilton' (*P.* Peony × *P.* Redstart), 1962. The ultimate goal in red Slippers was reached in this hybrid. Perhaps this was the spur which goaded hybridists into experiments with less shapely species and hybrids, typified by 'callosum' style vinicolour reds.

◁ *Paphiopedilum* Claire de Lune (*P.* Emerald × *P.* Alma Gavaert), 1927. The colour of the original species used to produce these green and white hybrids remained unchanged through several generations and varying size was the only distinction. Paphiopedilums are among the few orchids which have shown resistance to cloning — perhaps fortunately in this case.

▽ *Paphiopedilum* Delaina (*P. delenatii* × *P. victoria regina*, subspecies *chamberlainianum*), 1977. *P. delenatii* has propensity for giving pink base colour to hybrids. Its potential was not fully explored, possibly because of a preference for larger, shapely Slippers. This changed in the late twentieth century and work was resumed with a selection of these smaller species.

▷ *Phragmipedium* Grande (*Phrag. caudatum* × *Phrag. longifolium*), 1891. Phragmipedium hybrids have never attained the eminence of paphiopedilums and cross-pollination of the two genera proved difficult and unrewarding. Phragmipediums were Seden's province, if anyone's, but few were grown by the 1990s.

▽ *Dendrobium* Ainsworthii (*Den. aureum* × *Den. nobile*), 1874. The seed of this cross-pollination was sown in 1867 by the British hybridist Ainsworth. The hybrid provided the background for nearly all the *Dendrobium nobile* type of cross-pollinations.

◁ *Dendrobium* Montrose (*Den*. Ainsworthii × *Den*. Thwaitesiae), 1927. The parentage of *Den*. Thwaitesiae included the mysterious yellow-gold species *Dendrobium signatum*, an illustration and revision of which was given in Growing Orchids Book Three, *Vandas, Dendrobiums and Others*. There were thus three infusions of yellow to gold included in the pedigree of *Dendrobium* Montrose.

▽ *Dendrobium* Kobayashi (*Den. nobile* × *Den*. Merlin), 1968. In the pedigree, *Dendrobium* Merlin is a third-generation hybrid. Although the cross-pollination of *Den*. Kobayashi was made early in the twentieth century, it was not named until it was used by a Japanese hybridist many years later. It is described more fully in the text (page 56).

△*Dendrobium* Sunburst (*Den.* Merlin × *Den.* Thwaitesiae). The inbreeding common to soft-cane dendrobiums is expressed as much by this grex as any other. The cross-pollination was made in the 1920 decade, probably at the time of the *Den.* Kobayashi combination. More information is available in Growing Orchids Book Three, *Vandas, Dendrobiums and Others*, which features the genus.

△*Dendrobium* Malones (*Den.* Akatuki × *Den.* Glorious Rainbow), 1973. These hybrids completely changed appreciation of soft-cane dendrobiums. The *Dendrobium* Malones grex carried the best elements of Yamamoto-style flowers into their blend of shape and colour.

◁*Dendrobium* York (*Den. regium* × *Den. nobile*), 1963. *Dendrobium regium* appeared in hybridising much later than other soft-cane dendrobiums. Its history and an illustration also appear in Growing Orchids Book Three, *Vandas, Dendrobiums and Others*. It brought in the clear-coloured, inner labellum, a feature of some Yamamoto hybrid soft-canes.

◁ *Dendrobium* Oriental Paradise (*Den.* Evening Glow × *Den.* Oborozuki), 1971. The initial infusions of the yellow-to-gold original species appear again in this hybrid, which probably included the albino form of *Dendrobium nobile.*

▽ *Dendrobium* Sagarik (*Den.* Amethyst × *Den.* Lady Hamilton), 1960. These hard-cane dendrobiums have developed greatly in the last fifty years, but there was a sameness, except for shape and size, after initial cross-pollinations.

▷ *Dendrobium* Superstar (*Den.* Malones × *Den.* Utopia,) 1978. If a pattern had to be set to to judge the requirements for shape and colour in combination, *Dendrobium* Superstar could be used to exemplify it.

▽ *Dendrobium bigibbum* and its affiliates were responsible for a series of hard-cane hybrids which almost dominated warm-climate cultivation of the genus in the twentieth century. They were elaborated into a number of subsidiary cross-pollinations, but the white to purple-red hybrids generated from the complex remained unchallengeable.

△ *Dendrobium* Jan Orinstein (*Den. primulinum* × *Den. pierardii*), 1979. Completely different from the *Dendrobium nobile* type, but floriferous, attractive, winter dormant and more tolerant of cool temperatures than some soft-cane dendrobiums. These Indo-Asian hybrids are uncommon in collections and should be grown open-rooted, with some allowed to become pendant.

△▷ *Dendrobium* Sri Racha (*Den.* White Gem × *Den.* Theodore Takaguchi), 1982. The albino form of *Dendrobium bigibbum* and its sub-species, *Den. schroederianum* and *Den. phalaenopsis*, also played their part in what would be almost a purple-red monochrome of hard-cane dendrobiums without their influence.

△ *Dendrobium* (*unnamed*) (*Den.* Cheunson Rex × *Den.* Sarakai). As in most cultivated hybrids, the requirement is always for a rounded outline with a good profile. Cross-pollinations such as this probably reached the full potential of hard-canes and it remained only for growers to practise their skill in cultivating them.

△ *Dendrobium* Gatton Sunray (*Den. dalhousieanum* × *Den.* Illustre), 1919. This hard-cane hybrid is insular and incompatible with nearly all other hard-cane dendrobiums. It is inbred, difficult to grow for most cultivators and infrequently seen. When suited it is an admirable dendrobium in every way.

▷ *Dendrobium* Farmeri-thyrsiflorum, a natural as well as horticultural hybrid, quite distinct from other hard-cane combinations. These Indo-Asian species have been in cultivation for most of the history of dendrobium growing. They should flower after dormancy on mature growth of the previous season, with some from the same growth in more than one season.

◁ *Dendrobium* Esme Poulton (*Den. bigibbum* × *Den. kingianum*), 1978. A beautiful Australian indigenous combination, yet rare in collections. Although the capsule parent is tropical and the other an intermediate-climate dendrobium, it is possible to grow the hybrid in a moderate climate and flower it in autumn to early winter.

▽ *Dendrobium* Rosemary Jupp (*Den. striolatum* × *Den. teretifolium*), 1975. I refer to this hybrid, raised by V. and N. Jupp, in the discussion of the use of Australian dendrobiums with strange morphology in the text. The Australian catalogue of hybrids has enlarged considerably in the last fifty years, thanks to the work of many cross-pollinators, but principally of Walter Upton of Sydney.

△ *Masdevallia* Angel Frost (*Mas. veitchiana* × *Mas. strobelii*), 1982. In the ninety-odd years between the registration of *Mas.* Stella and that of *Mas.* Angel Frost, there were few entries. The genus regained popularity in the late twentieth century, but was not widely cross-pollinated, probably because of diminutive sexual parts.

◁ *Masdevallia* Stella (*Mas. coccinea* × *Mas. estradae*), 1890. The cross-pollination registered by the British grower Hincks who, although he had several early entries to his name, did not appear in most early records of growers.

△ *Masdevallia* (unnamed) (*Mas. maculata* × *Mas. ignea*). A masdevallia hybrid generated in the 1970s. Many of these have remained unnamed, despite their attractive flowers. If hybridists make cross-pollinations and market them, it is their responsibility to register them.

▷ *Masdevallia* (unnamed) (*Mas. sulcata* × *Mas. guttulata*). The 350 to 400 masdevallia species and associated genera have been hybridised more lately than in their earlier history, which began in 1794 with Ruiz and Pavon's discovery of *Masdevallia uniflora* in Peru.

◁ *Masdevallia* Machu Picchu (*Mas. coccinea* × *Mas. ayabacana*). This hybrid was registered in 1986, a product of Bealls, an American orchid nursery which raised a number of cross-pollinations in the 1970s and 1980s. There is considerable attraction in these small, floriferous, primary hybrids, most of which prefer intermediate climates and moist environments.

▷ *Masdevallia* (unnamed) (*Mas. peristeria* × *Mas. veitchiana*). Whenever *Mas. veitchiana* is used as either pollen parent or capsule carrier it appears to dominate and most other masdevallias cannot give equal size or colour. Are hybridists looking for novelty or advances in selective breeding?

▷ *Masdevallia* (unnamed) (*Mas. coccinea* × *Mas. xanthina*). Hybrid masdevallias are more incidental to orchid growing than some models produced by other genera. It is as much subject to 'fashion' as all orchids and has the disadvantage of being noncommercial.

△ *Odontoglossum* × Elegans (*Odm. cirrhosum* × *Odm. hallii*). This natural hybrid, referred to in the text, was discovered in the late 1870s. Speculation was frequent and divergent about the affinity of these species, but it appears that they are related. *Odm. cirrhosum* was illustrated in Growing Orchids, *Expanding Your Orchid Collection*.

◁ *Odontoglossum* Stroperry (*Odm.* Perryanum × *Odm.* Stropheon). This typifies the 'ultimate' odontoglossum, with conformity to required standards for size, shape and colour.

▷ *Odontoglossum* × Wilckeanum (*Odm. crispum* × *Odm. luteo-purpureum*). A natural hybrid. The dividing line between 'true' species and numerous natural hybrids in the genus was difficult to define in the nineteenth century. (Joan Skilbeck reproduction)

▷*Odontoglossum* × Elegans. Pollitt's plant, described in the text on pages 144–5. This photograph is of a lithograph.

▽▷*Odontoglossum* × Harvengtense (*Odm. crispum* × *Odm. triumphans*). In a genus much exposed to cross-pollination, this natural hybrid had forty-six cross-pollinations by 1945 and many were taken further, into second and third generations, in that period. Nevertheless it was overtaken by *Odm. eximium* (with 179 cross-pollinations) and *Odm. crispum* (with 260). (Reproduced by Joan Skilbeck from a 1903 photograph.)

▷ *Odontioda* Pele (*Oda*. Elac × *Oda*. Aviemore). Accentuated red with the coloured patterning in reverse. In one cross-pollination it is possible to obtain very different flowers, each with its own attraction.

▽ *Odontioda* Albert Park (*Oda*. Fremar × *Oda*. Pumari). When the cochlioda was introduced to odontoglossum hybridising it brought red intensity, at times, and different patterning. The first hybrid between an odontoglossum and cochlioda, *Odontioda* Vuylstekeae, flowered in 1904. (It is illustrated in Growing Orchids Book Two, *The Cattleyas and Other Epiphytes*.)

△ *Vuylstekeara* Edna (*Miltonioda* Harwoodii × *Oda.* Charlesworthii), 1921. Brilliant colour from two infusions of cochlioda, however small the resultant hybrid. The species *Odm. harryanum* and *Cochlioda noetzliana* were responsible for *Oda. Charlesworthii*, in a complex mix.

◁ *Odontonia* Floramusa (registration uncertain). Odontoglossum × Miltonia brought an addition to the complex which gave a broader, flatter labellum. The number of registrations is small compared with those for odontoglossums and odontiodas.

△ *Wilsonara* Tigersette (*Oncidium tigrinum* × *Oda*. Carisette). A variation in cross-pollinating which led to the creation of the composite named wilsonara. In this hybrid the oncidium colour influence was washed out by the odontioda.

◁ *Odontocidium* Artur Elle (*Oncidium tigrinum* × *Odm*. Hambuhren Gold), 1981. The golden-yellow of oncidium influenced the grex in a cross-pollination from the West German nursery of Artur Elle, although the majority of odontocidiums were produced by American sources.

▷*Odontoglossum bictoniense* × *Epidendrum gracile*. An unnamed grex from the late twentieth century, giving an unusual combination. The odontoglossum has been renamed in the genus lemboglossum, with the epidendrum a Bahamas and Cuban native and perhaps rightly classified as encyclia.

▽*Maclellanara* Pagan Lovesong (*Odontocidium* Tiger Butter × *Brassia verrucosa*), 1980. This plant further extends the affiliation of the genera into brassias. The resultant hybrid flowers were influenced principally by brassia.

△& ▷ *Odontioda* Wimmera (*Oda*. Matanda × *Oda*. Haniesto), 1982. An Australian hybrid bred by G. McCraith. Two photographs illustrating the difference in flowers from the same clone in consecutive years, as noted in the text.

▽*Phalaenopsis* Helen Smoothey (*Phal*. Freed's Angel × *Phal*. Freed's Curvaceous), 1979. Depending on their perception, growers were attracted by these striped forms. However the penchant of some judges for clear colour sometimes directed attention elsewhere.

△ *Phalaenopsis* (unknown) (*Phal.* Zauberose × *Phal.* Harlequin). The capsule parent was a West German derived grex. The genus usually proved easy to hybridise, adapted well to seed culture methods and grew quickly to flowering size. A number of hybridisers duplicated results of other pollinators. From seed to flowering could take as little as three years.

◁ *Phalaenopsis* Doris (*Phal.* Elizabethae × *Phal.* Katherine Siegwart), 1940. A grex which set a pattern to be followed into the inter-generic field. A pure white phalaenopsis with a white background which passed its good points on for half the twentieth century, when registrations surpassed those of any other genus. Most hybridising occurred in America and Hawaii.

△ *Phalaenopsis* (unknown) (*Phal.* Abram McCandless × *Phal.* Cindy Tsai). This grex was unregistered at the time this book was published. When the flow of registrations is high the number of unregistered hybrids is also high, frequently causing confusing entries on plant labels.

◁ *Phalaenopsis* (unknown) (*Phal.* Jack Haggard × *Phal.* Stripe Me Pink). As with pure white flowers, the aim was rounded shape and gaps filled between petals and sepals. The flowers are easily converted to these requirements because of their natural shape.

▷ *Phalaenopsis* Incognito × *Phalaenopsis* Malibu Carni. The symmetry and rounded outline of this unnamed hybrid typify the desires of growers.

△ *Doritaenopsis* (unknown) (*Dtps.* Pueblo Jewel × *Phal.* Lois Jansen). *Doritis pulcherrima*, a small Asian orchid, similar in many ways to phalaenopsis, brought brilliant red-purple to the complex, also blending white to pink in delicate shades. Doritis was not naturally well shaped like phalaenopsis, and cross-pollinators had to work on their hybrids to bring them into line.

▽ *Phalaenopsis* Astrolabe (*Phal.* Mildred Karleen × *Phal.* Artigny), 1976. The inclusion of a red labellum broke the sequence of white phalaenopsis and the addition of light striping altered the views of growers and most judges. The introduction of smaller species with colour advantages created new classes, occasionally smaller but much brighter.

△ *Doritaenopsis* (unknown) (*Dtps.* Clarence Schubert × *Phal.* Anne Marie Beard). This hybrid carried rich, deep colour into the combination. This colour was not repetitive and had to be periodically renewed or cross-pollinating had to be done with colour retention in view.

▷ *Miltonia* Anne Warne (*Milt. bluntii* × *Milt. spectabilis*), 1949. The name originated at Warne's nursery in Hawaii. Warne was the man who recommenced cross-pollination of the Brazilian section of the genus after World War 2. Moir, of the same islands, also shared this work on hybrids.

◁ *Miltonia* Bleuana (*M. vexillaria* × *M. roezlii*). The beginning and, if the truth was told, almost the end of a most fascinating elaboration of the genus into a rainbow of colour which transcended any expectations Alfred Bleu may have had when he cross-pollinated the two species in 1883. But these summer-flowering orchids were never as popular as other genera.

▷ *Miltonia* Jean Sabourin (*Milt.* Aurora × *Milt.* Piccadilly), 1956. It is difficult to imagine that the intensity of colour in this hybrid could have been generated from selective breeding of the original species, which were principally Colombian and light coloured.

▷ *Miltonia* Baden-Baden (*Milt.* Robert Paterson × *Milt.* Kasserine), 1959. The patterning of the labellum, occasionally referred to as the 'waterfall' effect, brought variation to the hybrids.

◁ *Miltonia* Connie Warne (*Milt.* × Castanea × *Milt.* Anne Warne), 1971. Brilliance in the genus is not confined to the Colombian miltoniopsis but is equally notable in Brazilian hybrid miltonias. The combination of miltonia and miltoniopsis did not produce what one would expect.

▽ *Miltonia* (unnamed) (*Milt.* Golden Fleece × *Milt.* Guanabara). Unlike the Colombian miltoniopsis, Brazilian miltonias may have long racemes with the flowers spread over their length. This unnamed hybrid has *Miltonia bluntii* as a progenitor and carries few flowers.

▷ *Miltassia* Charles M. Fitch (*Brassia verrucosa* × *Miltonia spectabilis*), 1961. The two genera combine to produce startling flowers, most of which have the spidery form of the brassia.

▽ *Oncidium* Nuuanu (*Onc. marshallianum* × *Onc. sphacelatum*), 1948. A typical oncidium hybrid in the form so well known to cultivators. These warm-growing orchids produce flowers in late summer to autumn and are spectacular when grown open-rooted into large plants.

◁ *Oncidium (unnamed) (Onc. alata × Onc. altissimum).* In common with many genera, oncidiums became subjects for what is best termed 'experimental' hybridising. Oncidiums generally did not have the driven exploitation toward 'form' which affected commercial genera such as cymbidiums or 'cattleyas'.

▽ *Oncidium* (unnamed) (*Onc.* Susan Perreira × *Onc. guianense*). A separate section of the genus named the equitant group, principally of Caribbean and north-east South American origin. Cross-pollinations into third and fourth generation hybrids have been developed to carry more flowers and colour than the small number on species from which they originated.

◁ *Oncidium* Arila (*Onc.* Palmyre × *Onc.* Sultamyre), 1981. Part of the inbred line of hybrids developed from *Oncidium varicosum*, a Brazilian species with widespread habitats in the southern states of that country.

△ *Brassidium* Black Beauty. This untraceable hybrid between brassia and oncidium has much to commend it to cultivators in appearance alone. The name could be false, but one can only look at labels, research them and take them at face value.

▷ *Vanda* Miss Joaquim (*V. hookeriana* × *V. teres*), 1893. This alliance of two terete vanda species was the first recorded hybrid in the genus. It was still grown as a garden plant in tropical climates in the 1990s.

△ *Vanda* (*Euanthe*) *sanderiana*. This native of the Philippine islands is the single most important species for hybridising vandas, indelibly stamping a pattern of colour and reticulation on the lower sepals of its descendants, however long the line and whatever the involvement of other species.

▷ *Vanda* Rothschildiana (*V. coerulea* × *V. sanderiana*), 1931. This plant was imported into Australia , probably from one of the first cross-pollinations. The reticulated blue species *Vanda coerulea* needed extensive work on sibling cross-pollination to approach the shape of *Vanda sanderiana* and bring the hybrid to perfection.

▷*Opsisanda* (unnamed). Quite common in warm climates, this grex is from the combination of vandopsis and vanda. It has probably remained unnamed because of flower shape and what to some is an unattractive colour.

▷ *Vanda* Arnothai (*V*. Lenavat × *V*. Onomea), 1974. The background of *V. sanderiana* is obvious after a number of generations, with the pink upper petals developed from the coloured varieties of that species.

△ *Renanthera* Brookie Chandler (*R. monochica* × *R. storiei*), 1950. Related to vandas, this brilliant hybrid was combined with a number of other genera. Withstanding almost direct sunlight and a tropical or semitropical climate, it was seldom grown satisfactorily in cooler climates, even with adequate heating arrangements.

◁ *Ascocenda* Yip Sum Wah (*Vanda* Pukele × *Ascocentrum curvifolium*), 1965. Although an early arrival in the alliance between these two genera, this hybrid proved outstanding and held its place until late in the twentieth century. The vanda used in the combination was a direct *Vanda sanderiana* derivative.

△*Renanthopsis* Lena Rowold (*Renanthera storiei* × *Vandopsis lissochiloides*), 1948. Part of the post-war diversity, when cross-pollinations in almost every cultivated species became the order of the day. The renanthera-vandopsis composite gave hard-growing plants suited to subtropical and tropical cultivation.

▷*Doricentrum* Pulcherrimin (*Doritis pulcherrima* × *Ascocentrum miniatum*). The alliance of a phalaenopsis affiliate with an ascocentrum may seem to be miscegenation, but if it is subject to further cross-pollination the end of such a hybrid line is an open book. This hybrid was produced by the University of Hawaii Horticultural Department in 1971.

△◁ *Ascocenda* Teoline Symphony (*Ascocenda* Yip Sum Wah × *Vanda* Rothschildiana), 1970. The blue *Vanda coerulea* is predominant and could not be suppressed. The quality of *Ascocenda* Yip Sum Wah as a seed parent is also obvious. A candid submission could be put that small ascocendas add nothing to vandaceous orchids except miniaturism.

◁ *Mokara* Bibi (*Arachnis hookeriana* × *Ascocenda* Pimpa), 1976. A trigeneric combination which always gave colour but shape in only few instances. Asian and Hawaiian nurseries produced their best hybrids in the late twentieth century, a period of expansion in mixed vandaceous pollinations such as the mokaras.

◁ *Vascostylis* Tham Yuen Hae (*Rhyncovanda* Blue Angel × *Ascocenda* Ophelia), 1968. The combination of rhyncostylis, vanda and ascocentrum. The colour is from *Vanda coerulea* and *Rhyncostylis coelestis*, both from Thailand.

▷ *Sartylis* Blue Nob (*Sarcochilus hartmannii* × *Rhyncostylis retusa*), 1973. An Australian cross-pollination raised by W. and J. Cannons. These hybridists also produced the sarconopsis (sarcochilis × phalaenopsis), featured in Growing Orchids Book Three, *Vandas, Dendrobiums and Others*. Sartylis was one of the first combinations of a vandaceous orchid with sarcochilus.

▽ *Aranda* Noorah Alsagoff (*Arachnis hookeriana* × *Vanda* Dawn Nishimura), 1940. In the middle years of the twentieth century only two intergeneric hybrids from arachnis were listed and other intergeneric associations were also few. In the following fifty years the list grew to hundreds, with arachnis involved in more than forty alliances.

△*Calanthe* Bryan, syn. William Murray (*Calanthe vestita* × *Calanthe vestita* var. *williamsii*), 1894. Raised and named by Cookson, who specialised in calanthes in the late nineteenth century. Nomenclature has been at the mercy of so many taxonomists that margins of credibility for some calanthes are unclear.

▷*Angraecum* Veitchii (*Angraecum eburneum* × *Angraecum sesquipedale*), 1899. This was the only hybrid raised in the genus until 1960. Like most primary hybrids, *Angraecum* Veitchii grows strongly and flowers consistently. It was only in the last twenty-five years that further cross-pollinations were made in the genus and its affiliates.

◁ *Phaiocalanthe* Irrorata (*Calanthe vestita* × *Phaius grandifolius*), 1867. There were few hybrids in any genus which preceded this James Dominy cross-pollination. It appears in his list of hybrids in the text (pages 8–9) and was registered by the Veitch nursery.

◁ *Phaiocymbidium* Chardwarense (*Cym. giganteum* × *Phaius grandifolius*), 1902. This frequently disputed hybrid could become the first orchid to be tested genetically to prove its status as an intergeneric cross-pollination. It begs that attention.

▽ *Catasetum* Orchidglade (*Ctsm. pileatum* × *Ctsm. expansum*), 1974. Few hybrids have arisen from this genus because of the habit of producing male and female flowers so irregularly. This grex was made with a self-pollinated clone, according to the label, so the habit persists into secondary hybrids.

◁ *Epiphronitis* Veitchii (*Epidendrum radicans* × *Sophronitis grandiflora*), 1890. A cross-pollination which quenched the epidendrum urge to grow tall and reduced this miniature to about fifteen to twenty centimetres high. The hybrid was a Veitch production and the cross-pollination was probably made by one of the nursery staff.

▽ *Sophronitella violacea.* A Brazilian cool-growing orchid which, since its discovery in the early nineteenth century, has evaded the arts of cross-pollinators except for one unsubstantiated instance in the early twentieth century, recorded in the text (pages 114–115).

△*Cymbidiella pardalina* syn. *rodocheila*. This Madagascan species which has proved difficult to self-pollinate or cross-pollinate is referred to in the text (page 184).

▷*Stanhopea* (unnamed) (*Stanhopea grandiflora* × *Stanhopea bicornuta*). This genus should be regarded as an open-root type. As the flower buds travel downwards, there should be scope for penetration of material so they can emerge pendant to open. Solid containers defeat this and should not be used.

◁*Pleione* Versailles (*P. formosana* × *P. limprichtii*), 1966. Pleiones may be grown at lower temperatures than most orchids. Indo-Asian, they occur at high altitudes, where they behave as deciduous plants which flower in late winter to spring in the young shoots, mature the pseudobulbs and enter winter dormancy.

◁*Zygopetalum* B. G. White (Z. Blackii × *Z. mackayi*), 1957. The cross-pollinators, using darker species and hybrids, have produced two or three generations of well-marked blooms, the flower spikes of which are seldom as tall as those of *Z. mackayi*.

◁*Zygopetalum* Helen Ku (*Z. maxillare* × Z. Blackii), 1967. Hybridism in the genus was begun in about the last quarter of the nineteenth century. The hybrids fell into disuse as a glasshouse plant. Few grexes were produced through the following years until almost 100 years later, when the albino *Zygopetalum* Helen Ku won many admirers.

The so-called Yamamoto dendrobiums were not bred solely by Yamamoto, but followed on from European hybridists and, in a minor way, from British and American hybridists and growers with a fondness for soft-cane dendrobiums. In some instances the Yamamoto-type hybrid flower stem is shorter than on earlier hybrids and the growths or pseudo-bulbs are occasionally shorter and stouter. The flowers, because of their size and shorter stems, appear to cluster more tightly, but rarely droop.

Whatever vision hybridists had in their minds for cross-pollinations this was possibly the ultimate of which the species' gene pool was capable. Miscegenation has frequently taken place in the genus as a whole and one must speculate whether these Yamamoto-type hybrids have been subject to the processes applied to other genera.

Before leaving the soft-cane section we should consider the apparent lack of success in combining the hard-cane section with it. The challenge has consistently been taken up by hybridists but apparently they have not persisted with it long enough. There have been some successful efforts, but most prove sterile in the primary hybrid stage.

However, it is still possible that a catalyst for combining soft-cane and hard-cane types does exist and a combination of the two might exceed in beauty anything that appeared in the middle to late twentieth century. This result would lead to the abandonment of more than 100 years of work and the appearance of a completely new look in the product.

It is impossible to make a complete survey of the soft-cane complex, but a few others are among illustrations in the third book of the Growing Orchids series. The main and important section is the one based on *Dendrobium nobile* and a very few associates.

There is, however, a large 'in-between' section not entered in this book. Its appearance in general culture is so insignificant that it should not displace the main themes, although several interesting cross-pollinations between what should be regarded as hard-cane and soft-cane dendrobiums are worth notice.

The first of these was between *Dendrobium moniliforme* and *Dendrobium pulchellum* (syn. *D. dalhousieanum*), raised in the Veitch nursery and named *Dendrobium* Stratius. It was flowered in 1892 and first published under the name 'Stratium.'

A second hybrid from *Dendrobium pulchellum* was obtained using *Dendrobium nobile* to carry the seed capsule. Although exhibited by Sir Trevor Lawrence, the plant was raised in the collection of another grower, who was immortalised by having the grex named after him as *Dendrobium* Whitei in 1900. The flower was large like that of *Dendrobium pulchellum*, but it had much of the colour of *Dendrobium nobile*. It received an Award of Merit. The grex was first named *Dendrobium* Dalhou-nobile but not so registered.

These two hybrids exhibit the compatibility between the two types to a degree, but it is significant that the line ended there and justified the term 'mule' for hybrids.

It becomes more difficult to draw a line between the two types of dendrobium as time and cross-pollinations add to the increasing list of hybrids generated in the last half of the twentieth century. Fashion has long since ceased to affect the composite genus and it becomes a matter of production setting the pace and perhaps pandering to the 'I want a piece of that' mentality.

Dendrobium growers are no less prone to the habit of discarding plants than other orchid growers. The contrast between the number of grexes cultivated and the number appearing in the seven volumes of *Sander's List of Orchid Hybrids* is an indictment of their bad habits.

Hard-Cane Dendrobiums

The name of these plants is self-explanatory and, compared with the often lush appearance of their relatives, the hard-cane dendrobiums are well described. They should be considered quite separately from the soft-cane dendrobiums.

While the soft-cane group is principally Indo-Asian, most hard-stem dendrobiums originate at almost the opposite end of the habitat and, while there is a mixture of the two types throughout, they could be regarded as South-East Asian to Australasian. Many are based on the phalaenanthe group and the overall appearance is of rose-purple flowers carried on slender stems arising or pendant from the head of pseudo-bulbs. The flowering season, particularly of the hybrids, is spread over much of the year, unlike soft-cane hybrids and species, which flower principally in spring.

If recorded history may be relied on, the first hard-cane hybrid was between *Dendrobium speciosum* and *Dendrobium kingianum*, two Australian orchids. It was noted in the *Gardeners' Chronicle* in 1892 and registered as *Dendrobium specio-kingianum* in 1896. It was earlier named *Dendrobium* × *Spyersii* by Sir Trevor Lawrence, the hybridist responsible for the cross-pollination, but not registered as such.

The first recorded hybrid using *Dendrobium bigibbum* was said to be from the cross-pollination of that orchid with *Dendrobium linawianum* (now known as *Dendrobium moniliforme*), the capsule carried by *Dendrobium bigibbum*. The grex was named *Dendrobium* × Sybil. It was raised by Norman Cookson in Britain and exhibited at a meeting of the Royal Horticultural Society in May 1893. A publication of the time stated: 'The plant, which seemed to bear little trace of *D. bigibbum*, received an Award of Merit'. The description accompanying the entry, however, noted that the flowers had large, deep purple sepals and petals, which certainly did not come from *Dendrobium linawianum*.

As an authentic hybrid it represented a breakthrough between soft- and hard-cane types. But as with so many early cross-pollinations, it seems that it was incorrectly catalogued and the same hybrid name is registered in the soft-cane section with the parentage given as *Dendrobium crassinode* × *Dendrobium linawianum*. *Sander's List of Orchid Hybrids* is the only accredited source and we must discard the notion that the hybrid is from *Dendrobium bigibbum* and search for another 'first'.

Even without the benefit of Javanese-Malaysian records from 1938 to 1948, which are missing, a choice between several nurseries in the Java-Malaysian area offers evidence about the use of the phalaenanthe section. It is certain that these nurseries were on the point of producing a type which was interrupted by World War 2, but which became common in the second half of the twentieth century.

Before we search through the phalaenanthe dendrobiums for progenitors, a hard-cane hybrid developed toward the end of the nineteenth century should be considered. It is the cross-pollination between *Dendrobium chrysotoxum* and *Dendrobium pulchellum*, named *Dendrobium* Illustre, of Indo-Asian derivation. *Dendrobium* Illustre was one of the first combinations of hard-stemmed dendrobiums, a Veitch production, with *Dendrobium chrysotoxum* as the capsule bearer. When it first flowered it was very small and, after the fashion of primary hybrids, easily produced its two pale yellow blooms.

Sir Jeremiah Colman, an ardent dendrobium fancier, used *Dendrobium* Illustre in a back-cross with *Dendrobium dalhousianum* (the older name for *Dendrobium pulchellum*) early in the twentieth century. The resultant grex from the cross-pollination flowered and was registered in 1919 as *Dendrobium* Gatton Sunray. An illustration of it appears in this volume (also in Growing Orchids Book Three, *Vandas, Dendrobiums and Others*), flowering on a plant in my collection in the 1950s.

The point to be made here is that although this dendrobium was a magnificent grower, with pseudo-bulbs about a metre high and very stout, it did not flower easily and proved resistant to hybridising, unlike its parent *Dendrobium* Illustre. There is little doubt that Veitch and others tried using *Dendrobium* Illustre often enough but without success. In the following years, almost to the close of the twentieth century, no further additions have been made to that line except for the back-cross again to *Dendrobium chrysotoxum* with the pollen of *Dendrobium* Gatton Sunray. The resultant grex was named *Dendrobium* Golden Swan in 1956. Inbreeding is not always a good line to follow.

Early in the twentieth century another hybrid emerged from the combination of *Dendrobium dalhousianum* with *Dendrobium thyrsiflorum*, which again introduced two Indo-Asian hard-caned dendrobiums to a list becoming overburdened with hybrids based on *Dendrobium nobile*. This new arrival was named *Dendrobium* Triumph by the British grower Armstrong. It is significant that the line also seemed to end there.

Other hard-stemmed dendrobium species were used in the early years of the twentieth century, but most were outshone by flowers of the phalaenanthe type, inaugurated with *Dendrobiums bigibbum, phalaenopsis* and *schroederianum*, which appear worth considering as an entity with several varieties. Those are the names under which they appear in *Sander's List of Orchid Hybrids*, however they are dealt with in other literature.

The nature of these plants and the climates in which they originated gave little reason for cultivating them in Britain and Europe in the later nineteenth century. They occupied minor places in collections compared with soft-cane types, although even these orchids were only grown in numbers by a few ardent collectors.

In the late twentieth century the phalaenanthe line transformed hard-stemmed dendrobium growing in subtropical to tropical climates into a galaxy of white, pink and red-purple flowers. It probably started with *Dendrobium* Boisseyense, the cross-pollination of *Dendrobium bigibbum* and *Dendrobium schroederianum*, flowered and registered by Maison Henri Vacherot-Lecoufle, of France, in 1926.

The phalaenanthe complex originates in the Cape York region of Australia, and the islands to the north of it, with undefined limits. Although brought into cultivation in the late nineteenth century, *Dendrobium bigibbum* was originally found and named in the 1850s. Taxonomists have had a picnic with it ever since, but that has little significance in *Sander's List of Orchid Hybrids*. The fact that the three orchids have been used in a criss-cross network of thousands of entries does not help to straighten out the tangle.

It may be hard to imagine that dendrobiums declined in popularity so much that in the 1930s only four hybrids were recorded in three years. There were very few in the later part of that decade, too. The generation of hard-cane types moved from Britain and Europe to the Asian region, principally to the Javanese islands. Bandoeng and Jakarta, in Western Java, were two centres which developed more hard-stemmed dendrobiums at that time than any other region.

Interesting entries occur in the hybrid list for what could be assumed as the period immediately before World War 2. They concern the nomination of Nagrok as registrant of three hybrids, which are *Dendrobium* Caesar (*D. phalaenopsis* × *D. stratiotes*), *Dendrobium* Albertine (*D. bigibbum* × *D. dalbertisii*) and *Dendrobium* Australia (*D. bigibbum* × *D. toftii*). These three entries are credited to Nagrok with no further identification. The responsible cross-pollinator could have been Nagrok Orchideenkweekerij, Soekaboemi, Java.

Dendrobium Bangkok (*D. phalaenopsis* × *D. taurinum*) was another obvious Asian hybrid. The registrant, Dang Toi Nursery, has no address or date attached, although it is possibly also Javanese.

Dendrobium Pauline, produced by Stuart Low in Britain in 1932 from the cross-pollination of *Dendrobium phalaenopsis* and *Dendrobium undulatum*, was an outstanding grex because it was the only hard-cane hybrid. There were few soft-cane hybrids in any of the lists of the 1930 period, but a principal registrant was Prince Shimadzu, of Japan, who became significant in the development of the hard-cane group.

Another entry in *Sander's List of Orchid Hybrids*, this time by the Australian Wilhelm Schmidt of Brisbane, gives a clue to the origin of some partly hard-cane dendrobium hybrids. He named *Dendrobium* Batavia in 1935, the cross-pollination of *Dendrobium* Statterianum and *Dendrobium taurinum*. In 1938 two plants of this cross-pollination grew in my glass-house, originating in Buitenzorg Botanic Gardens, Java. The plants never flowered and the cross-pollination was suspect because *Dendrobium* × *Statterianum* is a soft-caned natural hybrid.

Wilhelm Schmidt also registered *Dendrobium* Ellen (*D. kingianum* × *D. tetragonum*) in 1928, probably the second hybrid between Australian hard-caned species. The grex has been recreated many times since and grown, flowered and exhibited as magnificent plants in Australia and elsewhere. I met Schmidt in about 1942, when I was en route to the New Guinea war zone, and part of his library came into my possession in 1976.

Other hybridists in tropical locations contributed in minor ways to registrations, but it was not until the last half of the twentieth century that hard-caned dendrobiums were brought into prominence as orchids to be grown in all areas, with suitable heating arrangements to maintain tropical environments where necessary.

Putting aside other hard-cane cross-pollinations which had minor effects on the breed in general, the start of a line still being pursued towards the end of the twentieth century occurred in 1934. The year was notable for the registration of *Dendrobium* Orchidwood (*D. bigibbum* × *D. phalaenopsis*) by Orchidwood Inc., of New York, USA.

As much as any other, this grex was used with derivatives from *Dendrobium* Sanders Crimson (*D. schroederianum* × *D. taurinum*) and other hybrids to produce orchids which had to succeed. *Dendrobium taurinum*, a Philippines species, is usually referred to as an 'antelope' type, with long, dark, red-purple, twisted petals and red-purple labellum, features modified by phalaenanthe dominance.

Phalaenanthe-type hybrid dendrobiums are typically shaped like their progenitors. The flowers have a distinct resemblance to those of the genus phalaenopsis. The graceful arching racemes have been developed to the extent that they are longer, the flowers are larger and rounder and come in all colours from white to pink, rose-purple, blue-purple and, in extremes, tinged with a darkness verging on black. There is no need to nominate grexes because very few early hybrids, such as *Dendrobium* Lady Constance (1947), were discarded for purposes other than lack of space or a wish to keep 'toeing the line'.

While the natural flowering period of phalaenanthe-type species and associated hybrids is autumn to winter, the plants have been modified to the degree where they lack flowers only in summer, the growing season of these dendrobiums. In artificial cultivation there is little pause in growth during an annual cycle.

In considering hard-cane hybrids there are other lines which should not be overlooked. One developed in Australia associates a group of Queensland orchids, including *Dendrobium canaliculatum*, *D. discolor*, *D. johannis*, *D. antennatum*, *D.* × *Superbiens* (the natural hybrid between *D. bigibbum* and *D. discolor* syn. *D. undulatum*), and *D. trilamellatum*. These hybrids are not cultivated in numbers equal to those of the phalaenanthe group. Fashion, or perhaps size, has influence here.

In addition to that series, other Australian hard-caned dendrobiums have been cross-pollinated into a diverse section which, although not morphologically or florally large, is important to Australian growers, some of whom have devoted almost total collections to them.

These cross-pollinations, which include species such as *Dendrobium linguiforme* or *Dendrobium striolatum,* could be regarded as very curious if the morphology of those two Australian dendrobiums is understood. One of the hybrids is illustrated on page 98.

That section is almost beyond the scope of this book. It is dealt with more completely in Growing Orchids Book Four, *The Australasian Families.*

Masdevallias

Masdevallias were among the first orchids in the Americas to attract attention. But they have never been widely cultivated in species or hybrid form. A rather brief resumé of their history and cultivation was given in Growing Orchids Book Three, *Vandas, Dendrobiums and Others*. Enthusiasm for the complex — the best term to describe the 350-odd members of the genus — has ebbed and flowed over almost the whole of the nineteenth and twentieth centuries. They have never been an important part of the pastime, in fashion for some years and then dropping out of favour until something, such as new and brilliant flowers, brings them back again.

The masdevallia genus is native to South America, principally to the north-western states of Colombia, Ecuador, Peru, Bolivia and, to a lesser extent, Venezuela, growing on the eastern slopes of the Andes, with a few species in Central America and Brazil. This large genus is broken down into subsections or subspecies, with the whole regarded as part of the pleurothallideae and consisting of about 10 per cent of that group.

The altitudes of natural habitats range from just above sea level to about 3000 metres (approximately 10 000 feet). This difference means that masdevallias cannot all be grown in one environment, which is what most beginners attempt, and frequently fail, to achieve. This should also apply to hybrid masdevallias, but there is always a blending factor to take into account and some hybrids combine more easily than species.

Although the species have been cultivated for almost 200 years, hybridism in masdevallias has failed to reach the intensity of other genera such as the cattleya complex or the monotypic cymbidiums and paphiopedilums. It would be difficult to find more than four generations, whereas the comparative figures for cymbidiums and paphiopedilums would be perhaps four to five times that number.

For a considerable period few species were progenitors of hybrids, but in the last half of the twentieth century it was common to find curious combinations which gave rise to new grexes. Most species or hybrids, however, were thoughtfully linked, principally using the outstanding *Masdevallia veitchiana*, which was brought to England and Europe early in the last half of the nineteenth century and is still highly regarded. The brilliant orange of this masdevallia, with its ever-changing purple-red papillae, remains dominant in most instances, even when combined with much darker coloured species.

One of the first hybrids was *Masdevallia* Chelsoni, the result of cross-pollination of *Masdevallia amabilis* and *Masdevallia veitchiana* (the pollen parent). It flowered in 1880 and was noted for brilliant colour and the way the tint of the papillae encrusting the surface of the blooms changed with the direction of the light falling on them. This is one of the principal attractions of the pollen parent. The hybrid was named by Veitch, the registrant.

The reverse cross-pollination was made simultaneously and flowered at about the same time. It was given the name *Masdevallia* Splendens, which ultimately became its varietal tag.

The second registered hybrid was the combination of *Masdevallia ignea* and *Masdevallia coccinea*. It was named *Masdevallia* Fraseri in 1882 after Fraser, a newcomer to the ranks of the cross-pollinating fraternity, who was noted in the *Gardener's Chronicle*. This hybridising fraternity was far from popular in its time with the 'senior citizens' of the orchid fanciers, who preferred species orchids and found their 'pollution' beyond forgiveness.

Masdevallia coccinea was brought into cultivation in the 1840s and most growers would recognise its brilliance more readily than that of *Masdevallia veitchiana*. It seems that only four, or perhaps a few more, masdevallia species were known and grown in the first half of the nineteenth century and not many more appeared until the 1880s and 1890s.

The third hybrid, registered in 1884, was from *Masdevallia davisii*, again using *Masdevallia veitchiana* as the pollen parent. The resultant grex was named *Masdevallia* Gairiana, described as a brilliant orange-yellow with violet papillae. *M. davisii*, of course, is a cadmium yellow, deeper coloured in some clones. Seden was the hybridist and Veitch the registrant.

Then came *Masdevallia* Glaphyrantha in 1886, from the cross-pollination of *Masdevallia infracta*, a type that flowered more than once on the stem and often produced more than a single flower. The hybridist combined it with *Masdevallia barlaeana*, a single-flowered type on which the stems were produced anew annually

Masdevallia infracta, a Brazilian orchid, found and named in 1835, was included in *Paxton's Botanical Dictionary*, published in 1868. It is most variable and has been subdivided into a group of subspecies. It is caespitose, whereas *Masdevallia barlaeana* has what could be described as a creeping rhizome.

The fact that it would be difficult to find these hybrids in any collection in the 1990s points up the frequency with which fashion changes and is another example of the loss of many early hybrid grexes.

Another significantly different hybrid was generated from the cross-pollination of *Masdevallia tovarensis* and *Masdevallia ignea*, which produced two or three ochre-yellow flowers to a stem, opening consecutively. The pure white *Masdevallia tovarensis* belongs to the group on which the flowering stems persist for a number of years, two or three flowers opening together on a stem in the plants' annual cycle. Mature *Masdevallia igneas* produce single flowers only once. This hybrid was named *Masdevallia* Hincksiana in 1887.

Other hybrids were produced in the late nineteenth and early twentieth centuries, but until the 1950s there were few growers and only forty-five hybrid masdevallias had been registered, principally by British growers. This was a poor record of cultivation compared with, say, the cattleya complex. Indeed, only five cross-pollinations were registered in the first half of the twentieth century.

Natural hybrids occurred in the genus and the one between *Masdevallia barlaeana* and *Masdevallia veitchiana* was noteworthy. It bore two names and the two grexes were supposed by Reichenbach to have been from separate parents. Veitch later proved them to be one by a check cross-pollination and he named it *Masdevallia* Splendida. Reichenbach named his 'find' *Masdevallia* Parlatoreana. A later paragraph elaborates the history of these two masdevallias.

It was not until the 1970s that hybridists again began to cross-pollinate species and some hybrids. While many of these used *Masdevallia veitchiana*, other species also figured in pollinations, but not to the extent that their potential warranted. Although *Masdevallia veitchiana* has several outstanding cultivars, including cross-pollinations of different grexes, it appears to be dominated by the cultivar or variety *Prince de Galle*.

Compared with the hybrids generated by this species, cross-pollinations of other species seem to have novelty rather than competitive qualities. Only when grown into specimen-type plants with multiple flowers do they offer the spectacle one expects from hybrids generated from *Masdevallia veitchiana* with only two or three flowers.

Considering the morphology of the flowers when compared with the coarse floral parts of such orchids as cattleyas or cymbidiums (using the word 'coarse' in its broadest sense), it is not surprising that few hybridists could use the genus for their purposes.

Writing in *Reichenbachia* at the end of the nineteenth century, Frederick Sander had this to say:

> Twenty years ago it was thought but little less than an impossibility to attempt the rearing of hybrid Masdevallias, to say nothing of other genera such as Phalaenopsis and Odontoglossums. But, as is ever usual, nature had led the way abroad where these glorious Orchids are wild, and so her handiworks have encouraged the hybridiser at home, who has solved so many difficulties in cross-breeding, and whose deepest faith is, and must ever be, that in the garden 'nothing is impossible!' . . . And so our hybridisers at home have at last made not only some of nature's 'old lamps' or wild species, into new ones, but they have made for us the old lamps themselves, doing synthetically what the botanist analytically had perceived or prophesied in times past.

(The spelling and capitalisation are that of the time.) Doubtless this extract, published in the fourth volume, referred to the illustrations of the hybrids *Masdevallia* Courtauldiana, *M.* Geleniana and *M.* Measuresiana, all beautifully reproduced but not one of which was generated from *Masdevallia veitchiana*.

Of the parents used in the late twentieth century, it appeared that diversification was the order of the day, although it is apparent that flowers larger and more colourful than the rest owe their existence to *Masdevallia veitchiana* and its derivatives.

Reichenbach, usually credited with being the most assiduous and knowledgeable of botanists dealing with orchids in the late nineteenth century, said of the hybrid *Masdevallia* Geleniana, registered by Sander in 1887 or 1888: 'This is the debut of Mr F. Sander in orchid hybridising and very many more are forthcoming.' The writer failed to foretell the tremendous part which was to be played by Frederick Sander and those assisting him as professional and noteworthy growers and breeders in the early twentieth century.

The two hybrids *Masdevallias* Splendida and Parlatoreana came to cultivation as wild plants, gathered in conjunction with a consignment of *Masdevallia veitchiana* and *Masdevallia barlaeana* from somewhere near Cuzco, in Peru.

In 1878 Reichenbach described a flower from *Masdevallia splendid*a as a 'new species', adding that it 'made one think of a mule between *Masdevallia veitchiana* and *Masdevallia barlaeana* or *amabilis*'. He inadvertently or mistakenly wrote the name as 'spectabilis' instead of 'splendida.'

He named *Masdevallia parlatoreana* a year or more later, indicating that he proposed *Masdevallias veitchiana* and *amabilis* as the possible parents. *Masdevallia amabilis*, however, did not grow in the same locality as *Masdevallia veitchiana* and was ruled out.

Seden proved the status of *Masdevallia parlatoreana* as a natural hybrid resulting from the cross-pollination of *Masdevallias veitchiana* and *barlaeana* by hand-pollinating the two parents and flowering the seedlings, proving that *Masdevallias splendida* and *parlatoreana* were one. The name *Masdevallia parlatoreana* was then lost.

Despite large numbers of species, the list of natural hybrids is comparatively small. There is a qualifying factor, of course: namely, that many could have gone unrecognised because, as thriving naturally occurring new grexes, they became so numerous in fulfilling their evolutionary role.

In the last half of the twentieth century interest in the genus was renewed by more frequent generation and presentation of hybrids, blending well-known progenitors with such unusual co-partners as *Masdevallia maculata* or *Masdevallia strobelii* and other species.

Masdevallia Angel Frost, from the combination of *Masdevallia veitchiana* and *Masdevallia strobelii*, produced flowers which took the best qualities of each species and elaborated them into beautiful orchids which have potential for cross-pollination.

Since the 1970s there has been sporadic use of the genus in more variety than in its previous history, but few hybridists were associated with the breeding programmes. These were dominated by two or three American growers.

It is significant that there was little or no apparent attempt to change the characteristics of traditional masdevallia shape; the focus was on colour, with no preferences. Most attributes sought came originally from the species *Masdevallia veitchiana*, followed perhaps by the hybrid *Masdevallia* Angel Frost and similar flowers.

Veitch played the greater part in cross-pollinating the genus at the end of the nineteenth century. He was, of course, a commercial grower whose objectives perhaps differed from those of the several private growers of that time.

Among these was Hincks, about whom little is recorded but who named several early cross-pollinations. Warner, one of the originators of the beautiful *Orchid Album*, which was published in sections in the late nineteenth century, was another notable grower of the genus. Hincks is listed in other literature as a captain, probably military but perhaps seafaring, who lived at Richmond, Yorkshire, in the last quarter of the nineteenth century and had a collection of the genus. He was guilty of indiscriminate cross-pollination without keeping valid, accurate records of his work, with the result that the parentage of *Masdevallia* Acis, registered in 1890, has been the subject of speculation rather than authority.

The specialist growers and botanists of the period who, perhaps from practice, were adept at this sort of conjecture, deduced that its parents were *Masdevallia veitchiana* combined with *Masdevallia abbreviata*. It is registered under this parentage, although the order is reversed, with *Masdevallia abbreviata* carrying the seed capsule and *Masdevallia veitchiana* contributing the pollen.

Captain Hincks was also responsible for *Masdevallia* Stella (*M. coccinea* × *M. estradae*). Both parents were Colombian species from similar intermediate environments. *Masdevallia* Stella survived the lapse of a century and is still cultivated in 1990, a beautiful product of the hybridist's art. The illustration which appears on page 99 may be incorrectly named, but that was the name appearing on the label.

Other prominent orchid growers of the period, although they did not individually make cross-pollinations and raise the seed, possibly contributed to the pastime and named the hybrids so created. These men included Sir Trevor Lawrence, president of the Royal Horticultural Society in the late nineteenth century, whose grower, Bickerstaffe, made several cross-pollinations and raised seed, one of which was *Masdevallia* Falcata (*M. coccinea* × *M. veitchiana*).

Indicative of the carelessness of cross-pollinations in the early years is the example of *Masdevallia* McVittiae (*M. coccinea* × *M. tovarensis*, 1892), which again posed problems for botanists because the flowers did not agree with what could be expected from the parentage. The Royal Horticultural Society gave it an Award of Merit and the plant was said to have been cultivated from seedling stage for about fourteen years before flowering. In the *Gardener's Chronicle* Rolfe declared that it bore a remarkable resemblance to *Masdevallia* Measuresiana, derived from *Masdevallia tovarensis* and *Masdevallia amabilis*. Considering the time it took the plant to flower, it appears from reports that the cross-pollination must have been made in the glasshouses of W. Thompson in the 1870s. It is illustrated in a very old sale catalogue for the Thompson estate which is in my possession.

In conclusion, it is interesting to note that almost alone in orchid species used in hybridising, *Masdevallia veitchiana* has maintained such isolated eminence that it has seldom been superseded by any hybrid masdevallia. Some of the fine varieties have been lost, but in a group of masdevallias I bought in the late 1980s there was a plant of this species which was the equal of any I ever owned.

Searching for a pattern in the cross-pollinations which took place toward the close of the twentieth century did not disclose very much. The principal hybrids in use were derived from *Masdevallia veitchiana* and that species appeared dominant. While size is a consideration, it is impossible to expect larger flowers to stem from the small species which produced many of the hybrids of that period.

The genus was divided into sections between 1970 and 1990 and it is from members of subsections like *Dracula chimaera* that further developments may come. A problem to be overcome in hybridising from a section with pendant flowers is the need for remedial cross-pollinations that will give some of the qualities of this type without its 'bad habits'.

The degree of assiduity needed to use the pollinia in the first place is worth mentioning because, fortunately for hybridists in other genera, none had to cope with the minute sexual parts of masdevallia flowers.

Odontoglossums

This genus has a long history of cultivation, presenting as many problems associated with polymorphic forms as other orchids. For example, *Odontoglossum crispum*, named by John Lindley in 1844 or thereabouts, is so diverse in form and colour that hybrid origin is almost beyond doubt. The time lapse from original species to its metamorphosis at the period of first collection and cultivation is unknown.

Several other species and natural hybrids were associated with *Odontoglossum crispum* in various habitats. The passage from one to another could be a matter of speculation until it became wearisome to take it any further. In the first and second volumes of the *Orchid Review*, in 1893 and 1894, the editor, Robert Allen Rolfe, took up the challenge and solved some problems, leaving few questions unanswered. In particular he dealt with natural cross-pollinations.

The first problem, according to Rolfe, came to light about 1867, in plants imported by Hugh Low and Co. Reichenbach, one of the world's busiest botanists working with orchids at the time, named one plant *Odontoglossum* × Andersonianum, surmising that its parents were *Odontoglossum crispum* and *Odm. gloriosum* (syn. *Odm. odoratum*), both Colombian species. In reality it was a new species, but it was known as a hybrid. This explanation has been dealt with elsewhere in the Growing Orchids series.

In the following years considerable confusion existed, principally because of poor understanding of differences between species and their varieties and differences between those varieties and the flowers resulting from natural cross-pollinations. Some of this confusion is expressed in the fact that *Odontoglossum andersonianum* was known as *Odontoglossum ruckerianum* by some authorities and not others, a system of dual nomenclature not confined solely to this genus.

Robert Allen Rolfe had this to say of habitats and species:

> A species of whatever kind has always a definite geographical area over which the individuals which compose it are distributed, more or less continuously or disconnectedly, according to circumstances. Taking this area as a whole, the individuals are generally very numerous and everywhere they bear so strong a resemblance to each other that they can generally be recognised as belonging to the same species. Moreover, the present distribution, however wide, *must be ascribed to gradual diffusion from some original birthplace.*' (my italics).

This may be satisfying to those accustomed to studying plants, particularly orchid plants, and classifying them according to personal viewpoints. But as orchids have been in cultivation for so short a time compared with their occupation of habitats, it is not surprising that it takes some time to correctly place natural hybrids and their parents.

The odontoglossum complex is full of such examples and it is only because of men such as Robert Allen Rolfe and the experimentalists that so many have been proved. Among the latter were most of the major cultivators in the late nineteenth century mentioned in Growing Orchids Book Two, *The Cattleyas and Other Epiphytes*.

Most of this large Central and South American composite family was brought to cultivation in Britain and Europe in the second half of the nineteenth century. Various species originally nominated as odontoglossums have been reassessed for most of the twentieth century, so that about half have been transferred to new classifications with new generic titles. Most retained their original specific names, although those which did not fit botanical distinctions or duplicated names in use, were given new definitions.

In examining early imports of odontoglossums, botanists and taxonomists were impressed by certain plants from collecting regions where species were often accompanied by one or more associates in the field. The flowers of plants which attracted attention appeared to contain elements common to more than one species found growing in contact. Those different types of flowers were fairly quickly recognised as natural hybrids, as outlined in an earlier paragraph. This is common in many genera and not confined to orchids. Some botanists dislike the use of the word 'type', except when applied to herbarium specimens, but that is a short-sighted and narrow view of common usage.

The flowers in the complex that were regarded as natural hybrids included *Odontoglossum wilckeanum*, named by Reichenbach in 1880. It was introduced by Low and Co. of Britain in 1878, in a parcel of *Odontoglossum crispum* from Colombia, and identified as a natural hybrid very quickly by those skilled in analysing morphological distinctions. The first flowering of this stranger, however, was not in Britain, but in a Belgian collection owned by Massange de Louvrex and named for his gardener, Wilcke.

The characteristics of flowers of *Odontoglossum wilckeanum* were thought to have stemmed from *Odontoglossum luteo-purpureum* and *Odontoglossum crispum*, which grew in close contact.

The status of various proposed natural hybrids was confirmed principally by manual cross-pollination in glasshouse conditions once the genus was established in cultivation. Among those carrying out verification were several growers in British collections, principal among whom was the Veitch nursery.

European cultivators were also working on manual cross-pollinations of various species and in 1890 the manual cross-pollination between *Odontoglossum crispum* and *Odontoglossum luteo-purpureum* bloomed in Baron Rothschild's glasshouses in Paris. This was the genesis of odontoglossum hybridising after early difficulty in cross-pollinations. Apparently only a few seedlings reached maturity, flowered and were named *Odontoglossum* Leroyanum, but they were later relegated to the natural hybrid name of *Odontoglossum* × Wilckeanum.

This cool-growing genus cannot be fully covered in a book of this scope, but it is possible to follow development to the stage where history repeats itself and the further involvement of other genera is unrewarding. The word 'odontoglossum' is used occasionally in the following account as the word 'cattleya' is used, to designate that composite group.

The second hand-pollination to flower appeared in 1891. It was originally intended to prove the natural hybrid *Odontoglossum excellens*. This was mistakenly identified by Reichenbach as the cross-pollination between *Odontoglossum pescatorei* and *Odontoglossum tripudians*. Other authorities identified it as between *Odontoglossum pescatorei* and *Odontoglossum triumphans* and this parentage was subsequently proved by comparing the natural hybrid and the artificially raised cross-pollination.

In 1896 a flowering of Norman Cookson's cross-pollination between the species *Odontoglossum crispum* and *Odontoglossum hallii* received a First Class Certificate from the Royal Horticultural Society. The certificate was issued to *Odontoglossum* Crispo-hallii but the registered name of the hybrid was changed subsequently to *Odontoglossum* Cooksoni.

Charles Vuylsteke, of Ghent, Belgium, outstripped production of other breeders in the following years. *Odontoglossum crispum* came to the fore as paramount choice for most cross-pollinators and the most cursory examination of material used in illustrations gives the impression that its importance was not overemphasised. Vuylsteke again proved the natural hybrid *Odontoglossum harvengtense* by cross-pollinating *Odontoglossum crispum* and *Odontoglossum triumphans*. It was first named *Odontoglossum* Loochristiense in 1898, but later reverted to the natural hybrid name *Odontoglossum* × Harvengtense, which was given to the new discovery in 1894. Parentage was thought to be *Odontoglossum crispum* × *Odontoglossum sceptrum*, but the second name was ultimately proved wrong. The plant which carried the seed capsule is uncertain, although *Sander's List of Orchid Hybrids* offers *Odontoglossum crispum*.

Vuylsteke's cross-pollination of *Odontoglossum crispum* and *Odontoglossum harryanum*, originally known as *Odontoglossum* Crispo-harryanum, the name inscribed on the First Class Certificate awarded to it at the Temple Show, was also successful in Antwerp. One of the varieties was described as white with violet-brown markings. The grex was later named *Odontoglossum* Spectabile in 1899 and known by other synonyms, none of which predated the 1899 registration. The 1890s could well be referred to as the developmental period of odontoglossum hybrids, not only in Britain, but also in Europe, particularly France and Belgium.

Sir Frederick Wigan exhibited a plant of *Odontoglossum* Spectabile under its synonym *Odontoglossum* Crispo-harryanum, in Britain in 1898, supposed to have been generated from the reverse cross-pollination and to have been bought from Sander. But Vuylsteke, originator of the grex *Odontoglossum*

Spectabile, could not explain how this had come about. Perhaps strange things occurred in the closing years of the nineteenth century, just as they do in those of the twentieth. It is also possible that it was a coincidental cross-pollination.

The Temple Show, held in the Temple Gardens on the Thames Embankment, was the event of the year for the Royal Horticultural Society in Britain. It usually took place in May, late spring, and was the introductory point for many hybrids in various spring-flowering genera. Sir Trevor Lawrence, the president, exhibited a considerable collection of species, among them large plants of odontoglossums.

In Growing Orchids Book Two, *The Cattleyas and Other Epiphytes,* I mentioned the number of odontoglossums imported into Britain and Europe. The figure for plants purchased in thousands and accommodated in 'cool' glasshouses beggars description and it was not surprising that among the imports combinations had occurred between those species which grew together in such profusion. Spyers, the grower for Sir Trevor Lawrence at Burford Lodge, Dorking, Surrey, was a most able cultivator and no doubt it was his duty to select or discard from the thousands of odontoglossum plants which passed through the establishment. There was little thought, in those heady days for collectors, that the abundance could ever end, therefore only the best varieties survived the cull.

Robert Warner, another British grower of this species, was reputed to have had as many as 12 000 plants of *Odontoglossum* Alexandrae (*Odontoglossum crispum*) at a time in his cool houses between 1870 and 1900. In these and other consignments strange flowers appeared, accountable only as cross-pollinations between the species collected.

In an earlier period, a supposed natural hybrid was identified and named by Reichenbach as *Odontoglossum* × Hinnus, as early in cultivation history as 1870. Reichenbach mistook its parents because he did not know they came from Ecuador. He related the hybrid to *Odontoglossum luteopurpureum* and *Odontoglossum gloriosum* (*Odm. odoratum*), from Colombia. It eventually proved to be from the species *Odontoglossum cirrhosum* and *Odontoglossum cristatum*, both Ecuadorean natives. Probably *Odontoglossum* × Elegans, illustrated on page 102, should take principal place among early natural hybrids. A plant in flower was so named in about 1879.

There was also another plant which may be referred to *Odontoglossum* × Elegans, listed and described by Reichenbach in the *Gardener's Chronicle* for 1881 as *Odontoglossum* × Marriottianum. It originally appeared in a batch of *Odontoglossum cirrhosum* in the nursery of G. Marriott at Edmonton near London, and was known only by this description. Reichenbach's quandary in identifying the plant is reflected in his words describing it as one of the most difficult things he had ever had, fit to make him despair of himself or of nature or of both. 'I have considered it for a month and feel as ignorant as before. I cannot guess its descent'. Another plant supposed to be from the same parentage also appeared and was named *Odontoglossum* × Victor. It apparently did not survive.

Had Reichenbach known the origin of the proposed parents he may have had a clearer perception of the whole subject. He asked: 'How would a hybrid between *O. hallii* and *O. crispum* come between masses of *O. cirrhosum*?' *Odm. crispum* is Colombian, the other two Ecuadorean.

The true *Odontoglossum* × Elegans (1897) came from the species *Odontoglossum cirrhosum* and *Odontoglossum hallii*. An illustration by Joan Skilbeck, taken from a lithograph of a cultivar, also appears on page 103, originally bought as *Odontoglossum hallii* by H. G. Pollett. Considered by the Royal Horticultural Society, it carried four racemes bearing fifty-four flowers and was awarded a First Class Certificate. Numerous other plants were found in subsequent importations and given varietal names.

That this late nineteenth-century discovery caused so much trouble to botanists indicates the lack of knowledge about species at the time, perhaps caused by the vast and confusing numbers of plants taken from the various habitats.

To follow Rolfe's summary of the natural hybrid complex could take some time and space, so it is best to concentrate on artificial cross-pollinations, experimental as well as designed to prove the existence and identification of natural hybrids between closely occurring species.

After 1910 only twenty artificial and proven natural hybrids were registered. But before then only some 800-odd odontoglossum hybrids had been added to the total, using the generic title in its correct sense, not for the overall complex.

Growers with only minor experience, as well as those who have cultivated odontoglossums for years, recognise them as the most variable genus imaginable in species or hybrid form, a feature which has been appreciated by perceptive growers throughout the period of their culture. Along with their natural beauty and elegance, it is the principal attraction in what could be termed their ultimate phase. They have the added advantage of endurance in cool environments and most grow and flower well in temperatures as low as 10 degrees Celsius (about 50 degrees Fahrenheit).

One could say that hybridising in the late twentieth century is a matter of going up and down in one spot, if it were not for the amazing colour patterns distributed so irregularly over the flower segments. It adds something new all the time to hybridists' capabilities. The patterns in flowers of the same clone may change from year to year.

There are detractions from this diversification and beauty, however, and one of them has been present since the genus first appeared in orchid collections. The natural habit of the flower stems is from semi-horizontal to almost totally pendant and the trend towards larger and heavier flowers has imposed limitations in breeding upright, self-supporting stems. That weakness is the main failing of the whole complex and few are capable of maintaining an erect natural stance.

The step from hybridising within the genus and towards extraneous pollinations apparently began with the combination of *Cochlioda noetzliana*, a Peruvian species with similar morphological characteristics and the advantage of red colour. The first was carried out by the Belgian Vuylsteke in the late nineteenth century when he impregnated *Cochlioda noetzliana* with pollen from *Odontoglossum pescatorei*. This culminated in the flowering and naming of *Odontioda* Vuylstekeae in 1904.

Thus began the long list associating odontoglossums with genera such as oncidium, ada, brassia, aspasia, miltonia, comparettia and rodriguezia and so on, into combinations of as many as five genera.

While giving due credit to hybridisers producing these new flowers, all with devised or new names, none superseded the original odontoglossums in beauty or attraction, except perhaps in size and shape.

Different spellings of the species name *Cochlioda noetzliana* or *noezliana* may be found in writing by different authorities, such as *The Orchid Stud Book*, by Rolfe and Hurst, which gives the spelling as noetzliana, or *Sander's List of Orchid Hybrids*, which omits the 't'.

Producers of follow-on hybrid strains from about 1910 included European and American hybridists, and the complex was extended into further generic combinations as orchid growing increased.

Although the production of odontiodas began almost a quarter of a century later than that of odontoglossum hybrids, they outnumbered the odontoglossums by 1945. Use of the combination name odontioda (odontoglossum × cochlioda) followed a pattern adopted by registration authorities.

Almost every grower of note (including owners as well as gardeners and nursery staff) produced a number of hybrids, but the Charlesworth nursery far outstripped others, producing nearly 400 of almost 1000 odontiodas which were registered before the end of 1945. This figure represented registrations, and did not include the significant number which failed to meet the standards of the time or were not registered because of irresponsibility.

It may be difficult to assess the effect of *Cochlioda noetzliana*, except to say that the pigmentation introduced intensified the part-red of certain odontoglossum species. It carried smaller numbers of shapely flowers and they also were smaller. However, the broad-petalled, rounded shape which became a demand of fanciers was undoubtedly imparted by both genera. Very early in their culture, flowers having this attribute were valued above those with narrow floral segments and a star-like, pointed outline, even when the flowers were circular over all.

Although fashion dictated the popularity of the complex, odontoglossums, using the term broadly, have consistently had a model to work from. Assessment has demanded a broadening of petals and sepals toward the rounded configuration sought in all orchid hybridising.

The odontoglossum labellum, frequently deficient in broad flatness, was influenced in cross-pollinations with miltonias, which have broad labellums. The names of some of the genus have been amended to miltoniopsis and it was hybrids generated from *Miltoniopsis warscewiczii*, *Miltoniopsis vexillaria* and *Miltoniopsis roezlii* which showed most effect when brought into the odontoglossum breeding lines. The new genus so created again carried an abridgement of the two names and was known as odontonia.

There was a price to pay in this diversification, of course, because miltoniopsis carry few flowers. However, this was a temporary setback and corrected breeding over a period brought the flower count to within reasonable distance of odontoglossum and odontioda numbers.

The first odontonia was produced in 1905 from the cross-pollination of *Miltonia warscewiczii* with *Odontoglossum crispum* and the hybrid named *Odontonia* Lairesseae. It was far from the ideal demanded by the standards of those days, but the labellum was expanded to good proportion and balance, considering the overall configuration of the flower.

The Belgian grower, A. de Lairesse, showed the plant at a Royal Horticultural Society orchid committee meeting in London and was given an Award of Merit. While the flower was considered insignificant, resembling the miltonia, it was enough of a breakthrough to offer promise.

The inflorescence was about 0.75 m long (about 28 inches) with five short branches and carrying nineteen flowers in all. By the time the orchid committee looked at it, two of the flowers had gone, no doubt headed towards further cross-pollinations.

It had little influence, however, as only four derivatives from it as a parent were ever registered. The colour overlaying the white ground and labellum was reddish purple. Rolfe, who described it in the *Orchid Review*, seemed surprised at the dominance of the pollen parent, which he understood to be the miltonia. This does not agree with other records, which list the miltonia as the capsule bearer.

The second registration was *Odontonia* Culiginosa, in 1906, by the breeder Vuylsteke, the Belgian who created a furore with his original showpiece odontioda. Neither of the two better endowed miltonias (miltoniopsis) was used. Instead, Vuylsteke used *Miltonia warscewiczii*, native of Central and South America. The pollen parent was *Odontoglossum nobile* (syn. *Odm. pescatorei*).

Charlesworth followed in 1907 with the third hybrid, *Odontonia* Elwoodii, but he used *Miltonia roezlii*, a Colombian orchid, in combination with *Odontoglossum cirrhosum* as the pollen parent. Both were illustrated in Growing Orchids Book Two, *The Cattleyas and Other Epiphytes*.

The cross-pollinated trio, odontoglossum, cochlioda and miltonia (miltoniopsis), was given the generic title vuylstekeara and became the generating trunk for branches into other cross-pollinations with minor genera. Although these bred interesting diversities by the 1990s, the work which appeared important to orchid growing was accomplished almost at the beginning of the century. Vuylstekearas became large and most colourful companions for earlier intergeneric dominators. They belonged to the period between 1910 and 1950, later to be combined with oncidiums, from which issued the odontocidiums, maclellanaras and a large section of what could be called interesting miscellaneous orchids, only a small number of which could be said to supersede odontiodas, odontonias and vuylstekearas.

Odontocidiums (odontoglossum × oncidium) brought in a different genus and their inclusion added lighter tints such as yellow and gold to the already broad array of colour.

In the last quarter of the twentieth century the whole odontoglossum complex assumed a popularity it had never previously experienced, with a large American element almost dominated by R. B. Dugger overriding the one-time English and European dominance. Smaller countries also shared in the hybridising, and the Australian, G. McCraith, added to the odontoglossum and intergeneric registrations.

Odontoglossums assumed a status impossible to imagine when they were first introduced. Unlike other orchid alliances, such as the cattleya complex, this cool-growing series of plants retained the capacity to grow and flower in moderate conditions.

Wilsonaras, the combination of cochlioda, odontoglossum and oncidium, came from a series of cross-pollinations culminating in the production of flowering plants in 1916. The first was *Wilsonara* Insignis, the product of cross-pollinating *Odontoglossum illustrissimum* with *Oncidioda* Charlesworthii, the last named the pollen parent. As a choice *Oncidioda* Charlesworthii was poor because of its size, which is indicated in an illustration in Growing Orchids Book Two, *The Cattleyas and Other Epiphytes*.

Wilsonaras went into eclipse in the middle of the twentieth century and it was not until the 1980s that they again proliferated.

The seal of approval — such as awards given by authorities like the Royal Horticultural Society or other responsible national institutions — does not always last for ever. When looking at records, particularly modern photographic records which date from the 1950s, the changes in standards and criteria in committees and other bodies become quite apparent.

Sometimes deprecatory remarks are made about earlier awards, or even those of contemporary adjudicators, but it should be remembered that breeding alters the flowers of multigeneric or monotypic hybrids to fit patterns imposed by an ever-changing group of judges. These bodies may think they have solved the riddle of the perfect flower. In looking at changes in fashions they should take warning not to be too optimistic. Those earlier flowers fulfilled the desires of their period and are as important as those of later cross-pollinations because they were the source material.

There is a return to basic appreciation all the time and, to a large degree, flowers of the late twentieth century, in whatever genus, show this in much the same way as those of the end of the nineteenth century.

There are many facets to the make-up of perfect 'odontoglossum' flowers, from the initial quality of the plant, the length of inflorescence and the distance between individual flowers to the length of the individual stems which hold the flowers on the racemes. Stems have always been an important consideration. Some see only colour — and perhaps a favourite colour — others see shape. It amounts to what one expects to see. But if and when hybridists become wise they learn to *see all the details that make the whole*. In essence, nothing has changed and nothing ever does except fashion.

Phalaenopsis

The history of this genus was briefly related in Growing Orchids Book Three, *Vandas, Dendrobiums and Others*. It is confined almost totally to the northern hemisphere, notably in the area from the Philippines to as far west as Burma and northern India and extending as far south as Australia, Papua New Guinea and Indonesia. Phalaenopsis were among the first orchids noted by Linnaeus when he began his part in the enormous task of simplifying the nomenclature of all the known plants in the world. Their numbers were increasing so much that the system being used became unwieldy and indistinct. Many plants, trees and herbs of all types are still being reviewed and renamed as a result of this initiating activity in the eighteenth century.

Only a small part of the phalaenopsis hybrid section is reviewed in this book and, considering the number of registrations in *Sander's List of Orchid Hybrids*, it may appear that the coverage given to the genus is disproportionate. As with miltoniopsis, phalaenopsis has had much inbreeding to produce the flowers currently appearing. Most of these flowers are white and with little distinction separating them.

History is probably correct in recording a phalaenopsis as the first proposed natural hybrid orchid noted in Britain. Its advent is given in Growing Orchids Book Three, *Vandas, Dendrobiums and Others*. The flower was recognised speculatively by Lindley about half-way through the nineteenth century and later confirmed when Seden manually cross-pollinated *Phalaenopsis aphrodite* and *Phalaenopsis rosea* for the Veitch nursery more than thirty-three years later. It is recorded in Seden's list of hybrids earlier in this book, with *Phalaenopsis rosea* still bearing Lindley's name of *Phalaenopsis intermedia*. *Phalaenopsis amabilis* is the correct name for *Phalaenopsis aphrodite*, but in this book the names attached to the parents of hybrids at the time of registration are used.

It is surprising that these plants were successfully cultivated in the cooler British climate, considering that they were collected in the most tropical of climates, but a couple of instances of plants in flower exhibited in Europe in the nineteenth century prove it. The packing and transport of the plants was meticulously carried out.

If attention is given to warmth and moisture, it is possible to grow phalaenopsis as easily as other hybrid orchids. But cross-pollination was slow to begin. Seden's hybrid flowered in 1886 and a second hybrid from the same nursery flowered in the following year. It resulted from cross-pollination of *Phalaenopsis schilleriana* and *Phalaenopsis amabilis* and was named *Phalaenopsis* Rothschildiana.

There was some discussion between authorities about this hybrid. Reichenbach suggested that it closely resembled *Phalaenopsis leucorrhoda*, but Veitch subsequently said that this was incorrect, naming the pollen parent of his hybrid as the Malayan form of *Phalaenopsis amabilis*. *Phalaenopsis* Rothschildiana and *P. leucorrhoda* have been entered as natural hybrids in Sander's first list and the species *Phalaenopsis aphrodite* and *Phalaenopsis amabilis* are regarded as conspecific by the authorities, even though they carried different names in the early hybrid lists. Both names are still common in the 1990s.

In the same year (1887), the third hybrid was produced at the Veitch nursery from the cross-pollination of *Phalaenopsis amabilis* and *Phalaenopsis violacea*. It was named *Phalaenopsis* Harriettae, after Harriet, the daughter of Erastus Corning of Albany, USA. This pollination introduced an advance in hybridising within the genus because the parents were from different sections.

Phalaenopsis amabilis has a known history of more than 250 years, although it has not been cultivated in the western world for that period. So far as British culture is concerned, it is about 150 years since this plant was first flowered. It is no exaggeration to say that it has been the most important influence on hybridising since that time. Robert Allen Rolfe relates that J. H. Schroder was awarded a Silver Banksian Medal by the Royal Horticultural Society in 1850 for his presentation of this novelty.

Since that time several names or synonyms have been applied to *Phalaenopsis amabilis*, including the Australian variant, which was successively known as *Phalaenopsis rimestadiana*, *Phal. rosenstromii*, *Phal. amabilis* var. *rosenstromii* and *Phal. amabilis* var. *papuana* in various publications through the years since its discovery in the Cape York rainforests. It was one of the last to be renamed.

Schlechter made what was probably the first sighting of this variant in Papua New Guinea on the Malia River below the Bismarck Ranges in 1908. He later found it in other areas. It could be the subject of much speculation that *Phal. amabilis* is found in so wide an area of the western Pacific, South China Sea and Philippine islands.

Few hybrid phalaenopsis were added to the list in the 1880s, but in this period the first of so many mishaps apparently occurred when parents were not recorded as cross-pollinations were made. *Phalaenopsis* Leda, for which the parentage was proposed as *Phalaenopsis amabilis* × *Phalaenopsis stuartiana*, was identified by some as a variant of *Phalaenopsis leucorrhoda*. The doubt still continues because the plant originated as a stray seedling in the Veitch nursery. The flowers of the last-named grex are distinctly different from *Phalaenopsis* Leda.

In the 1890s as few as sixteen hybrid phalaenopsis were added to the list, but other species were included in cross-pollinations, such as *Phalaenopsis equestris*, syn. *Phalaenopsis rosea*, *Phalaenopsis mannii* and *Phalaenopsis sanderiana*.

The white species, *Phal. amabilis* and *Phal. stuartiana*, exercised considerable influence on the genus throughout its hybrid history. Like other genera, the genus did not develop into an overpowering number of grexes until the second half of the twentieth century. The outstanding early hybrids were few and principally bred from these two species.

The colour spectrum was poorly represented in early hybrids, but these two species blended well into early cross-pollinations with the pink *Phalaenopsis schilleriana*. There was a large amount of repetitious hybridising in the white and pink sections until the requirements of shape, size and colour were finally met. It could be said with asperity (and truth) that there was little to choose between some hundreds of cross-pollinations and the route to the 'ultimate' phalaenopsis could have been much shorter.

Some idea of the increase in numbers in the second half of the twentieth century can be gained by comparing the 150-odd registered hybrids in the first edition of *Sander's List of Orchid Hybrids*, 1945, with the 'guesstimate' total of more than 4000 registered names in the 1981–1985 list.

It is difficult to select an outstanding grex from those 150-odd early hybrids because, as with most genera subjected to cross-pollination, some speculation was involved. It is much easier for hybridists working now with select material from several generations who do not need to trace grexes on which to work.

The primary hybrid *Phalaenopsis* Elizabethae is a substantive member to use in tracing early phalaenopsis seedlings. It was raised by Vacherot and Lecoufle in France and named in 1927. The parents were *Phalaenopsis amabilis* and *Phalaenopsis rimestadiana*. Although the two are conspecific and the inbreeding becomes obvious, the two species (as they were regarded at the time), are both white but have slightly different flower construction. Like *Cattleya* Edithiae, *Phalaenopsis* Elizabethae is a controlling influence in its genus. Both are exceptional improvements on the parents.

The year 1927 was important for phalaenopsis culture because it was then that Vacherot and Lecoufle succeeded in the search for a formula suitable for seed growing. They raised plants and flowered them in a fraction of the time taken previously, but the report in *The Orchid Review* was unclear about the method used.

Phalaenopsis Elizabethae presents a curious puzzle, because *Phalaenopsis* Gilles Gratiot, raised by Dr Jean Gratiot of France, is also registered, with the same parentage, in the 1945 *Sander's List of Orchid Hybrids*. This was amended in a later list to *Phalaenopsis aphrodite* × *Phalaenopsis rimestadiana*.

Phalaenopsis Katherine Siegwart resulted from cross-pollination of *Phalaenopsis amabilis* and *Phalaenopsis* Gilles Gratiot. *Phalaenopsis* Elizabethae was used as the capsule-carrying parent in the cross-pollination with *Phalaenopsis* Katherine Siegwart, at the Duke Farms in New Jersey, to produce the outstanding *Phalaenopsis* Doris, which is still regarded as a phalaenopsis landmark. An illustration of the grex appears in Growing Orchids Book Three, *Vandas, Dendrobiums and Others*.

From that series of combinations arose the basis for the superbly shaped white or coloured phalaenopsis hybrids currently available. The plants used were those which were once separate but later consolidated under the specific title *Phalaenopsis amabilis*. This orchid was used in its various forms more than thirty-five times, as a capsule or pollen parent, in the initial list of hybrids, which numbered 150-odd. In secondary cross-pollinations it dominated as a progenitor.

The introduction of *Phalaenopsis schilleriana* to selected white-based hybrids brought pink to the breeding lines and additional flower numbers where they were thought to be needed. Stronger stems adequately compensated for the increase in flower numbers and the stems of the ultimate production-line hybrids are double the diameter of their white or pink antecedents. Nevertheless, the extra weight must be supported and most spikes on modern plants need staking when grown for exhibition purposes.

Most cross-pollinations concentrated on these exhibition aspects. But there was always a search for new colours or colour patterns and these were limited in the white breeding lines stemming from *Phalaenopsis amabilis*, *Phalaenopsis stuartiana* and their derivatives.

New objectives demanded considerable hybridising effort, selective breeding and, finally, choosing combinations which would give the required shape and size to flowers. The ideal, as in other breeding, has always been to develop a rounded outline in the principal flower parts, with the petals meeting on the centre line of the flower, to the point of overlapping without distortion. At the same time the sepals had to be broadened so that they filled the natural gaps. In general, the labellum had never played a principal part in 'designing' the flower until the introduction of deeper colour and the extension of labellums into features of the grexes of the late twentieth century.

It was never possible to breed a complete line conforming to these principles throughout and as fast as objectives were achieved the system of appreciation changed to make the requirements more difficult to attain. However, conformation was such that acceptable standards were reached in clear-coloured, paler flowers, white and pink to rose, by the 1970s.

To seek deeper colours hybridisers returned to species considerably smaller than *Phalaenopsis amabilis* and, in doing so, imposed size limitations on the flowers. In *Sander's List of Orchid Hybrids* for 1981–1985, twenty-six species appear in the breeding tables. Many had appeared before, but this indicates the nature of the search for new 'breaks'. This concerned only phalaenopsis, the deepest coloured of which is probably *Phalaenopsis lueddemanniana*. Intergeneric cross-pollinations are another matter.

The transition from clear-coloured flowers to the extraordinarily diverse and beautiful veining and colouring now achieved is remarkable in orchid breeding. The genus was cultivated in subtropical and tropical climates in numbers comparable with cymbidiums or the 'cattleya' complex, but the quantity of seedlings raised was exceptional. The fact that they could be purchased at little cost and flowered in two or three years in subtropical and tropical climates brought scope for marketing the product among growers as numerous as those usually found only in major genera. Phalaenopsis became popular and, considering their beauty, they were worth taking up. Of course a climate of warmth, moisture and constant air flow over the plants and flowers had to be provided if it did not exist.

The introduction of new species broadened the scope for breeders and new combinations began to appear and be accepted — something which was denied them when size and shape had dominated thought so short a time before. Orchid growers are fickle and so, of course, is fashion.

The genus has allies in the Indo-Asian sphere as well, and these were introduced to give diversification and, in some instances, enhancement.

Phalaenanthe Intergenerics

An early influence on cross-pollination was *Doritis pulcherrima*, which has closer affiliations with phalaenopsis than other genera. The colour of the flowers, although variable, is in purple-red tonings in darker specimens, verging into pale pink and rose in others.

Doritaenopsis Asahi was the first hybrid arising from *Doritis pulcherrima*, which was the capsule-bearing plant pollinated by *Phalaenopsis lindeni*. The hybrid was registered in 1923 by Baron Toshito Awasaki of Japan. No further registrations followed until 1955, when the Rod McLellan Company of California placed another with the Royal Horticultural Society, followed by John Miller, also of California, in 1956. There were only six, all told, before 1960.

Then began a rapid cycle of development, with *Phalaenopsis* Doris and its derivatives filling principal roles. The enduring breeding line which was developed from *Phalaenopsis* Elizabethae was the paramount feature of the genus. *Phalaenopsis* Grace Palm should perhaps be included in the honours, but it did little more than carry on as a descendant from *Phalaenopsis* Elizabethae and *Phalaenopsis* Doris.

Most of this progress was in Asia and America, including Hawaii. It continued until a record number of the genus phalaenopsis and its affiliates had been registered in *Sander's List of Orchid Hybrids* towards the end of the twentieth century.

Entries in the list of doritaenopsis from the first registrations grew to 450-odd in the index of 1980–1985. The registrants comprised a changing list of names, but John Miller of California stayed with the breeding line from the time when *Doritis pulcherrima* first entered the scene. The results of his experimental work are basic to the success of intergeneric hybridising.

Other genera with affiliations in cross-pollinating with phalaenopsis include aerides, arachnis, ascocenda, neofinetia, renanthera, sarcochilus, vanda and vandopsis. The quality of progeny from those genera did not inspire breakthroughs into combinations such as those of doritis hybrids.

W. and J. Cannons made outstanding cross-pollinations in Australia between phalaenopsis and sarcochilus. Three outstanding grexes were *Sarconopsis* Macquarie Sunset, *Sarconopsis* Macquarie Lilac and *Sarconopsis* Jean Cannons. *Sarconopsis* Macquarie Sunset is illustrated in Growing Orchids Book Three, *Vandas, Dendrobiums and Others*.

These hybrids were derived from *Sarcochilus hartmannii*, the largest Australian member of that genus, which is illustrated in Growing Orchids Book Four, *The Australasian Families*. Its influence is marked in those three hybrids, although they are distinctively of phalaenopsis origin.

Further work with phalaenopsis background material and intergeneric hybrids appears promising, but with diversification into fresh fields the original intention of hybridising to produce size as well as quality and shape had to be temporarily abandoned. In doritis cross-pollinations like the one illustrated on page 111, fresh fields opened up and novelties with little background of phalaenopsis appeared.

By the 1990s the ultimate had been reached in phalaenopsis and their derivatives and further work aimed at developing labellum size, pattern or colour. What is the ultimate 'phalaenopsis' flower supposed to look like, if not the one illustrated on page 111?

Part of the answer to that question appears in a book entitled *Phalaenopsis Formosa*, published by the Asia Agri-Business Corporation in 1987, with about 190 pages of colour illustrations of the phalaenopsis complex. Its extraordinary variety rivals the development of almost all other genera and much is still in the experimental stage, with primary hybrids from about thirty species. When these are introduced and selectively bred from, there is an infinite variation of spotting, lining and barring which presages a brilliant future for the genus in the twenty-first century.

There is a real threat that the cattleya complex will enter a decline, displaced by this newer flush of beauty and grace. Plant for plant, phalaenopsis grow faster and flower more freely than cattleyas. However, one factor emerges plainly — the best of phalaenopsis, from the point of view of shape, remain the white derivatives from the progenitor *Phalaenopsis* Elizabethae. There is little doubt that this will be a permanent yardstick to measure the quality of the genus.

Miltoniopsis and Miltonias

These two are morphologically distinct, although they once made up a single genus, distinguished, perhaps, by the comparatively 'hard' appearance of miltonia plants compared with those presently catalogued as miltoniopsis. Miltonias are principally confined to Central America, Venezuela, Colombia, Ecuador and Peru. A good guide to their incidence is given in *Venezuelan Orchids*, Volume 6, by Dunsterville and Garay (Andre Deutsch, London, 1959).

Miltoniopsis

The name miltoniopsis is not 'modern' in the sense of being applied recently. It was used in European orchid literature in 1889 and 1890 to designate the hybrid *Miltonia* Bleuana on which a whole series of hybrids rests like an inverted pyramid. It is hard to understand how so few original species could generate such a galaxy of colour. Admittedly others were included later, but the total is still small compared with a genus such as cattleya.

A brief history of the finding and cultivation of miltonias (miltoniopsis) was given in Growing Orchids Book Two, *The Cattleyas and Other Epiphytes*. At first hybridising was attempted with odontoglossums and, although a large number of seedlings resulted, none lived long enough to flower. Cultivation problems were no less a hazard in the nineteenth century than they were 100 years later. Miltoniopsis are rarer at exhibitions than genera such as cattleyas, partly because they flower in summer, a season in which few exhibitions are held.

The first hybrid miltonia was known synonymously as *Miltoniopsis* Bleui or *Odontoglossum* Bleuanum in the periodicals of 1880 to 1900. The details published in Veitch's *Manual of Orchidaceous Plants*, which include an illustration, indicate that the flowers more closely resembled *Miltoniopsis roezlii*, the pollen parent, than *Miltoniopsis vexillaria*, which was possibly the principal source of the attractive red colouring in the flowers of the middle to late twentieth century.

Alfred Bleu, the French grower who successfully raised this first hybrid, was pleased, not to say elated, about his new flower, which was much admired. The cross-pollination was made in the summer of 1883, the capsule matured the following autumn and the seed was promptly sown.

(This meant simply scattering it on the surface of the material used to pot the parent plants.) The records show that several plants reached flowering size, all appearing similar to *Miltoniopsis vexillaria*. The flowers were more like the pollen parent, with a diversity of colour. The principal change in some was a large yellow blotch at the base of the labellum.

Veitch repeated the cross-pollination and it is understood that the nursery produced plants which flowered a year after those of Alfred Bleu. Hybrid orchids of any kind were not the principal cultivars in the last two decades of the nineteenth century, mostly because so few were raised and hybridists, Veitch among them, held onto their productions.

Jules Hye de Crom, of Ghent, Belgium, took the next step by back-crossing *Miltoniopsis* Bleuana with *Miltoniopsis vexillaria*. The resultant secondary hybrid was named *Miltoniopsis* Hyeana. Other hybridists did the same with *Miltonia roezlii*, thus producing *Miltonia* St Andre.

These secondary hybrids were subsequently bred back to the parents in a sequence which had only two species as origins, and this was the base of the whole miltonia complex for many years. It reached such a degree of inbreeding that there was repetition throughout, and in 1938 the Orchid Committee of the Royal Horticultural Society suspended registration of new hybrids from this source. The following note appeared in *Sander's List of Orchid Hybrids* in 1946:

> In 1938 we decided to discontinue the publication of direct Miltonia Hybrids. One reason is that Miltonia Hybrids from the same seed pod vary in a great degree and almost identical forms are produced from differently named parents. At least nine-tenths of Miltonia Hybrids are, in fact, forms of *Miltonia* Bleuana, the very large majority come from the two species, *Miltonia vexillaria* and *M. roezlii*. The time has come to consider these Miltonias as garden forms and to give a suitable name to any outstanding variety. Hybrids that may be raised from other species than *M. Roezlii* and *M. vexillaria* as also all bigeneric crosses, we shall list as heretofore.

Despite the inbreeding, there was considerable colour variation in the flowers of the grexes because, although *Miltonia vexillaria* is most frequently known as a pink species, some clones are deeper coloured.

Although producing smaller flowers, at least two other species which were added to the breeding table brought the hybrids back into consideration. *Sander's List of Orchid Hybrids* later included names as they were submitted for registration and accepted by the Orchid Committee.

Two unrecorded, or poorly recorded, hybrids were *Miltoniopsis* Lord Lambourne and *Miltoniopsis* Princess Margaret, both of major consequence. Others of equal importance in the list also appear to have unclear parentage, but the vast bulk of cross-pollinations registered in the 1946 edition of Sander's list are simple interbred hybrids using two species.

It is revealing for growers of the genus to look at the list of hybrids in Sander's register and put a few derivations on paper. It is significant that of about 250 registrations in the list to the end of 1945, after more than half a century of hybridising, only four cross-pollinations between what were then regarded as miltonias included the species *Miltoniopsis phalaenopsis.*

Of equal significance is the fact that only five species of what are now known as miltonias were used in hybridising over all. These were *Miltonia candida, M. clowesii, M. flavescens, M. regnellii* and *M. spectabilis*. Taxonomic changes have affected some of these.

Since that first half-century of work, *Miltoniopsis phalaenopsis* and *Miltoniopsis warscewiczii* have been used. Neither has size to recommend it, but correct breeding patterns should be able to compensate for that. In the hybrid list both genera, with applicable notation, are still regarded as miltonias.

Many of these notations of taxonomic changes in *Sander's List of Orchid Hybrids* may add further confusion. *Miltoniopsis warscewiczii* is now listed as a miltonioides but, even here, the issue is confused by the 1981–1985 miltonia list, when the synonymity of *Miltonia endresii* is considered.

Miltonias, using the title in a broad sense, interbreed with several other genera, some of which offer surprising compatibility. For example the genera ada, brassia, aspasia, cochlioda, odontoglossum oncidium, tricopilia and rodriguezia have been bonded with miltonias through cross-pollinations since the 1950s. The compatibility of these diverse orchids from the Americas reinforces a view that orchids, as parts of the world's flora, once had much closer relationships than they now seem to have. Orchids probably offer more scope for botanical investigation than any other series of related plants.

Miltoniopsis hybrids continue to be cross-pollinated and the ultimate flowers, although produced in the first half of the twentieth century, are principally elaborated from such simple beginnings as two species.

There is no other case in hybridising in which cross-pollinations from two original species have produced such remarkable colour variations — from the deep intense reds of the eighth generation *Miltonia* Jean Sabourin (illustrated on page 113) to pure whites or yellow flowers like *Miltonia* Bellingham × Yarrow Bay, illustrated in Growing Orchids Book Two, *The Cattleyas and Other Epiphytes*. Some allowance must be made for the colour influence of *Miltonia warscewiczii* and *Miltonia phalaenopsis*, introduced following the suspension of registrations, but the basis was already there for intensified colour in the original pansy-like flowers of the early cross-pollinations.

If, in the end, nothing is left to produce new colours, there is an old fall-back position which is to change conformation, particularly to round out the flower parts and give a flat profile. These characteristics have always been points in favour of the species, where there is little furling. The texture or substance of the flower parts is fixed from the originals and little can be added other than by selective breeding. Sometimes, though, something unwanted appears when hybridists chase elusive factors, such as distortion or genetic disturbance.

In tracing the history of the species Frederick Sander, writing in *Reichenbachia*, added to the story by quoting the extraordinary restrictions on Reichenbach which made his work of describing *Miltoniopsis vexillaria* very difficult:

> ... having a bloom lent to him by a friend after a promise was made to him under five heads, viz: not to show it to anybody else, not to speak much about it, not to take a drawing, not to have a photograph made, not to look oftener than three times at it.

That may sound incredible, but modern parallels are not unknown. Frederick Sander also discussed the areas from which plants had been taken, the varietal differences between them and the fact that his business had imported at least a million plants. The demand was apparently insatiable. From his description it is apparent that there was much colour variation and difference in size from district to district and that he had found it impossible to cross-pollinate the orchid with odontoglossums. Yet at that time *Miltoniopsis vexillaria* was catalogued as an odontoglossum. The plants are morphologically similar and so the pollination barriers were surmountable.

His note in *Reichenbachia* ends with the observation that the true miltonias come only from Brazil and no odontoglossums are found there. *Miltoniopsis vexillaria* was at one time widespread in Colombia but is now obtainable only in restricted quantities in restricted areas. The collectors Bowman, Klaboch and Lehmann were the principal gatherers and exporters in the last quarter of the nineteenth century. No natural hybrids with other species were ever found by these people.

Miltonias

The genus carrying this name comes from a different environment from that of its relative, miltoniopsis. The two are partly compatible in cross-pollination, although the differences between them do not promote much mixing.

Between 1950 and 1990 little or no cross-breeding occurred, although the species had been known and grown since the discovery of miltonias near Rio de Janeiro in the 1830s. In fact, they were among the early orchid imports into British and European glasshouses. Growers who have cultivated *Miltonia spectabilis* will understand how easily it is accommodated. Its colour variations, too, make it an attractive addition to any collection.

The Brazilian section of the genus has seventeen or more members, eight of which have been recognised as natural hybrids. The growth habit is similar throughout, with indications that the plants are epiphytic with rising rhizomes, better suited to slab culture than pots. (This characteristic has been noted in most cross-pollinations.)

Records of hybrids occur in the first lists, generated from miltonia species rather than miltoniopsis. In the initial period *Miltonia candida*, *M. clowesii*, *M. flavescens*, *M. regnellii* and *M. spectabilis* were used.

Miltonias candida and *spectabilis* were used to create the first odontonias, using miltonias with odontoglossums. They were *Odontonia* Cybele, from *Miltonia candida* in 1913 and *Odontonia Lucilia,* from *Miltonia spectabilis* in 1914. The pollen parent in both instances was *Odontoglossum cirrhosum.*

Miltonias were not fully exploited until the 1950s, because growers were preoccupied with other genera. Since then several hybridists have used them successfully within the genus and as intergeneric pollen or seed parents.

From the cross-pollination of *Miltonia bluntii* and *Miltonia spectabilis,* came breeding stock which produced *Miltonia* Anne Warne for the breeder J. Milton Warne of Honolulu. This grex has been consistently used to further the line since its introduction in 1949. It was among the earliest of hybrids using the genus, inheriting a great amount of red-purple from its parents.

Albino varieties of miltonia species have never had as much use as coloured varieties in breeding programmes, wherever they were produced. The monotony of pure albino plants is perhaps responsible, although in their species form they are always popular. Well-flowered white orchids, whether species or hybrid, always seem to attract attention.

In 1961 *Miltonia bluntii,* a natural hybrid between *Miltonia spectabilis* and *M. clowesii,* produced the hybrid *Miltonia* Crimson Crest for the hybridist W. W. G. Moir of Honolulu, from cross-pollination with *Miltoniopsis warscewiczii.* Moir, an outstanding intergeneric hybridist, did a considerable amount of work with most of the orchids that are compatible with miltonias. This was an early introduction of the genus miltoniopsis, which remains a miltonia in registrations.

Intergeneric hybridising with miltonia, among others, began early in the twentieth century, first with bigeneric cross-pollinations, then trigeneric, quadrigeneric, pentageneric and, finally, the combination of six genera into the one hybrid form.

The compatible genera are aspasia, brassia, cochlioda, miltonia, odontoglossum and oncidium. The interbreeding of these six resulted in the production of a singular generic combination given the name *Brilliandeara* Gary. This grex, a product of 1982, was raised by W. W. G. Moir from the cross-pollination of *Forgetara* Mexico and *Burrageara* Sambu River.

The hybrid forgetara is formed from the cross-pollinations of aspasia, brassia and miltonia. The name honours the Frenchman Louis Forget, a collector in South America, a brief history of whose life and activities is given in Growing Orchids Book Two, *The Cattleyas and Other Genera.*

Burragearas are the combination of four genera — miltonia, cochlioda, odontoglossum and oncidium. The first burrageara was raised and named by the British hybridists Black and Flory in 1927 and named in honour of Albert Burrage, a noted American grower. The raceme, which was long, with well-spaced flowers, resembled that of *Oncidium macranthum,* the representative of the genus used in the combination. The colour was a combination of the other species and the typical yellow-gold of oncidiums was almost entirely absent. The only other hybrid of the period to integrate four genera like this was the potinara.

At the time burrageras were produced, three trigeneric combinations appeared in *Sander's List of Orchid Hybrids* — charleswortharas, vuylstekearas and wilsonaras. It is possible to trace back through progenitors to find which miltonias were used in all the combinations. It is evident from this analysis that miltonias were more prominent than miltoniopsis and, in some ways, it is regrettable that the two are now distinct.

Miltonias and miltoniopsis both have distinctive characteristics and hybridists who wish to produce certain types will probably find that intermixtures do not follow similar lines to those which pollinations produce within each separate genus.

Charlesworthara Alpha, the combination of *Miltonia schroederiana, Cochlioda noetzliana* and *Oncidium macranthum,* appeared in 1919. It involved a miltonia, as the complex was then known, not a miltoniopsis. The flowers were scarlet and the name honoured a British orchid nursery. It was the first of a line which has not progressed very much since.

Colmanara Sir Jeremiah, a 1963 hybrid, is worth mentioning because it honours a distinguished British orchid grower in two ways — by using his surname as the generic definitive and his title and Christian name as the clone title. The first colmanara was a combination of *Odontoglossum bictoniense, Miltonia spectabilis* and *Oncidium varicosum.* It was produced by the Hawaiian hybridist W. W. G. Moir, in a period unfortunately overshadowed by cymbidiums and other more popular orchids.

Moir probably had more authority, in combining miltonias with their compatible associates, than any other plant breeder. In his description of the production of *Brilliandeara* Gary which appeared in the *Orchid Review* of April 1983, several points emerged for hybridists. Moir's pollination of *Forgetara* Mexico with *Burrageara* Sambu River was made in 1976. Moir stated that the product in flower was unimpressive, but the combination and successful flowering of a hybrid derived from six different genera would be the triumph of any orchid breeder's lifetime, whatever it looked like.

Moir's programme was based on the production of Vuylsteke's hybrid odontioda in 1904. It continued with the burrageara from Black and Flory in 1927, the miltonidium (miltonia × oncidium) from Mansell and Hatcher in 1936, the brapasia (brassia × aspasia) from his own cross-pollination in 1957, the forgetara (aspasia × brassia × miltonia), also from his own cross-pollination, in 1972 and culminated in the brilliandeara in 1982.

It would be remiss to leave J. Milton Warne of Honolulu out of the honour list of hybridists working on Central and South American genera — his contribution was perhaps as important as that of Moir.

In the context of hybridising, 'miltonias' have no equal when they are considered as one genus, rather than separated by botanists into the two (or more) sections they now occupy. Their compatibility with other genera was previously remarked upon, but on analysis they would probably be incompatible with each other. While the miltoniopsis of the twentieth century are largely the result of inbreeding, the miltonias are based on more numerous species and are obviously more fecund and prolifically hybrid aligned. They are, however, far fewer than such cross-pollinations as 'cattleyas'.

Philosophical consideration of the miltonia complex leads to all sorts of speculation about their past, present and future. Have we really reached the end of the road? Did the brilliandeara and other products of persistent hybridising appear long before their natural evolution? What could have been expected if the genera had been left undisturbed until they fulfilled their role? Interesting questions indeed.

Brassias

Brassias are important because of their affiliation with miltonias and other orchids listed as the producers of so many hybrids. The genus was among early orchid imports into Britain and Europe; Sir Joseph Banks, for example, brought *Brassia maculata* from Jamaica in 1806.

Other species were slowly added over the next half-century and, despite the appearance in collections of six or more from Central and South America, brassias remained in few hands and never achieved the numbers of other imported and grown genera.

Popularity had much to do with hybrid production and a surprising facet of brassia culture was a lack of cross-pollination within or outside the genus. In the first century of brassia cultivation, no hybrids were registered. The first recognition of a cross-pollination which included brassia was a trigeneric hybrid named *Sanderara* Alpha, from *Brassia lawrenceana* × *Odontioda* Grenadier, recorded in 1937. Although it may not have been the first trigeneric combination, it was nevertheless remarkable. It was raised by Mr Louis Sander at the nursery at Bruges, Belgium and bloomed on a small plant. The flower was small (only 38 mm long), considering the size of the parents. The sepals and petals were long and narrow, brownish buff with occasional green patches. The labellum was broad and yellow, marked with light brown spots and edged with light brown. The plant received a preliminary commendation from the Orchid Committee of the Royal Horticultural Society in Britain.

In 1948 Kirsch, of Hawaii, registered *Brassidium* Coronet (*Brassia brachiata* × *Oncidium anthocrene*, the first brassia-oncidium hybrid. Other entries in the 1946–1960 hybrid registration list included two brapasias (brassia × aspasia) and nine brassidiums (brassia × oncidium). Four of these combinations came from W. W. G. Moir, a prolific source of new material.

There was still no brassia combination. It was not until 1958 that other significant cross-pollinations began to appear. Again it was Moir who provided three miltassias, the cross-pollination of miltonia and brassia. They were *Miltassia* Flower Drum (*M. warscewiczii* × *Brassia verrucosa*), *Miltassia* Premier (*M. spectabilis* × *Brassia caudata*) and *Miltassia* Puakinikini (*M. regnellii* × *Brassia allenii*). *Brassia allenii* has been transferred to the genus ada but retains its registration title of brassia. It is illustrated in Growing Orchids, *Expanding Your Orchid Collection*.

It was not until 1961 that two cross-pollinations of brassias were registered, again from Moir: *Brassia* Chieftain (*B. longissima* × *B. verrucosa*), and Brassia Gold Threads (*Brassia caudata* × *Brassia longissima*). The surprising thing about these was they received no recognition as 'firsts' in the *Orchid Review* of the period.

The first secondary hybrid also appeared in the 1961–1970 volume of *Sander's List of Orchid Hybrids* from the cross-pollination of *Brassia* Edvah Loo (*B. longissima* × B. *gireoudiana*) and *Brassia gireoudiana*, a back-cross probably aimed at intensifying the colour from the pollen parent.

Brassia macrostachya, syn. *Brassia lanceana*, was first illustrated in Lindley's magnificent production *Sertum Orchidaceum* in 1836. It came from Venezuela and the Guianas on the north-east coast of South America, and is only one of a genus of more than twenty members noted for their spidery outlines. Perhaps it was this morphological departure from a rounded outline which banished all such species from 'hybridists' notebooks. The desire for flowers with a rounded outline was an important motivation for hybridists when they chose so many genera for the massive lists first compiled by Rolfe and Hurst for their *Orchid Stud Book* and later by Frederick Sander for his list of hybrids.

It was only after a partial modification of paphiopedilum, cattleya and cymbidium breeding between 1960 and 1990 that most of the miltassias and other beautiful combinations came into being. Their advent began at least a decade earlier with speculation about whether brassias and other previously disregarded hybrids had a future. This may have led to a departure from the closed patterns of the pre-war hybridists by the experimenting new generation which changed so many other things in the world. Whatever it was, it brought brassias into more than twenty intergeneric combinations with as many as five genera in the more diverse cross-pollinations. In almost every instance brassias proved dominant. The outline of nearly all registered grexes, developed from whatever combination, had long petals and sepals, and the rather inadequate labellum of brassias was given size and shape from other genera. The most obscurely developed of these hybrids was probably the brilliandeara, described in the miltonia section, which had six genera in its breeding.

What further development is possible after extending a genus into so much different cross-breeding? It depends on what suits fashion at the time: intensification of colour or modifying the shape of the flowers.

Although there are minor breaks from principal generic combinations using the oncidium-odontoglossum complex, the development of these obscure cross-pollinations does not offer either commercial or collector prospects; however orchid growers of the twenty-first century may well turn to species orchids because of their rarity. That has always been a drawcard.

Oncidiums

Considering that oncidiums were among the first orchids to be brought into cultivation it is remarkable that they were hardly hybridised at all until the early twentieth century.

Some attention was given to oncidiums as cultivated species early in the nineteenth century but, as Veitch remarks in his *Manual of Orchidaceous Plants*, so few were known to botanists that only five were used by Olaf Swartz to establish the genus in 1800. It took about another century for them to be recognised as a large group of plants. Amongst early naturalists and botanists to note prospective acquisitions of South American oncidiums was Alfred Russell Wallace (a brief resumé of his life is included in Growing Orchids Book Four, *The Australasian Families*). Wallace travelled extensively in the Amazon region and recorded his experiences in the book *Travels on the Amazon and Rio Negro.*

Several natural hybrids were recognised in the early stages of cultivation but there is insufficient evidence that they were verified by manual cross-pollinations. Hybrid character was probably recognised and verified by discerning botanists and amateurs who grew and exhibited the plants. More information on this genus is included in Growing Orchids Book Two, *The Cattleyas and Other Epiphytes.*

The following are listed natural hybrids, with their parentage:

- *Oncidium gardneri* (*Onc. dastyle* × *Onc. forbesii*);
- *Onc. haematochilum* (*Onc. lanceanum* × *Onc. luridum*);
- *Onc. illustre* (*Onc. leucochilum* × *Onc. maculatum*);
- *Onc. mantinii* (*Onc. forbesii* × *Onc. marshallianum*);
- *Onc. punctatum* (*Onc. forbesii* × *Onc. gardneri*);
- *Onc. stanleyi* (*Onc. curtum* × *Onc. marshallianum*);
- *Onc. wheatleyanum* (*Onc. crispum* × *Onc. dastyle*).

The first recorded and registered cross-pollination was made by Charlesworth in 1909, more than 100 years after the first of the genus were grown as cultivated plants. It was generated from the species *Oncidium lamelligerum* and *Oncidium tigrinum.* A second hybrid named *Oncidium* Janssenii was registered two years later by the same nursery, this time using *Oncidium forbesii* and *Oncidium tigrinum. Oncidium lamelligerum* is an Ecuadorean orchid, *Oncidium tigrinum* is from Mexico and *Oncidium forbesii* comes from the Brazilian Organ Mountains region.

Oncidiums are liberally spread over all the Central and South American states. Any idea of the true number of the genus is confused by alterations in its generic status. There are several distinctions or sections among the members and from time to time some are moved into other sections or genera. The relative abundance of apparent natural hybrids is the result of perhaps thousands of years of juxtaposed development, extending into genera other than oncidiums.

The morphologically distinct Caribbean equitant section of the genus provided nothing of interest to growers before the middle of the twentieth century. But W. W. G. Moir's development, from *Oncidium pulchellum* and *Oncidium triquetrum* in particular, produced basic material for other hybridists to work on. Admiral Bligh brought *Oncidium triquetrum* to British gardens from Jamaica in 1793.

Moir used these equitant oncidiums to produce attractive hybrids more suited to warmer environments than intermediate oncidium cross-pollinations usually prefer. He also used the genus with compatible genera such as brassias, odontoglossums and miltonias.

The oncidium and allied genera section of *Sander's List of Orchid Hybrids* for the last half of the twentieth century reads like a handbook of W. W. G. Moir's cross-pollinations. Other hybridists worked with oncidiums, but none with the intensity and understanding of Moir.

Oncidium varicosum proved a prepotent species in the twentieth century. Beginning in the 1920s with *Oncidium* Boissiense (*Onc. forbesii* × *Onc. varicosum*), the pollen parent was frequently referred to in early listings as the variety *rogersii*. Use of the varietal name has persisted, rightly or wrongly, throughout the century. This cultivar is frequently noted in society exhibitions and national shows, but it is probably true that no growers have ever seen a plant of the original clone *Oncidium varicosum* var. *rogersii*. I have certainly never seen any of these clones equal to the measurements given for the original cultivar (see Growing Orchids Book Two, *The Cattleyas and Other Epiphytes*, page 191). That is not to deny, however, that many fine varieties are still in cultivation.

This genus is so large that it is impossible to list all the notable hybrids in a book such as this. The hybrids are difficult to cultivate unless correct environments are designed for them. They are principally bred along the lines of the *Oncidium varicosum* type, and W. W. G. Moir seems to have found that in this group a small proportion react as seed parents and not incorrectly as pollen parents. Choice of this small number of prepotent clones for cross-pollination is rewarding. It has probably never been proved whether there is a factor which could take over and reverse the roles of pollen and seed parent.

There are two options in oncidium breeding. The first is to breed for display with the large numbers of flowers generated by such species as *Oncidium varicosum, Oncidium sphacelatum, Oncidium flexuosum, Oncidium incurvum* or others with branching racemes. The second is to breed for size. It should be remembered in this context that it is almost inevitable that larger flowers need stronger stems and these will carry fewer flowers.

A third level is attained with the equitant type, which are suited to tropical and subtropical climates. None is cool growing like some of the derivatives of the first or second choices. The equitant-bred orchids usually carry more diversity of colour than those derived from typical Brazilian and associated South American oncidiums, which carry more yellow-to-brown tones. Strong rules in this regard, however, should not be binding or accepted without question.

There are several species with large and substantial leaves and almost no pseudo-bulbs, and from these it is impossible to generate many flowers. They are unsatisfactory as plants for intermediate culture and their hybrids should be crossed off the shopping lists by cool-climate cultivators, unless adequate heating and light is available.

Although the genus has been known and cultivated almost as long as any other, it has never reached the degree of hybridising demonstrated in such species as laelias and cattleyas from the American region, or the elaborate cross-pollinating of Indo-Asian genera such as dendrobiums or paphiopedilums.

Oncidiums have, however, been associated with several other genera in hybrids involving six genera, as in *Brilliandeara* Gary. This number may be increased, since various oncidiums are compatible with as many as sixteen or more genera according to records of cross-pollinating factors in *Sander's List of Orchid Hybrids*. Genera capable of integration into hybrids with oncidiums, as indicated in the 1985 Sander's list, include ada, aspasia, brassia, cochlioda, comparettia, gomesa, leochilus, lockhartia, macradenia, miltonia, miltoniopsis, notylia, odontoglossum, rodriguezia, trichocentrum and tricopilia.

The work involved in integrating all of these into a flowering combination is daunting, perhaps impossible. However, work already done with the larger flowered genera — such as odontoglossums, miltoniopsis and miltonias — in combination with brassias provides a background of trigeneric and quadrigeneric hybrids that should encourage more use in the future.

How worthwhile such proliferation might be depends on many imponderable factors, such as the correct sequence in which the species should be used in cross-pollinating or the correct combinations to use as final cross-pollinating grexes. These grexes would enlarge the scope for hybridists to concentrate on a series that could be useful innovations in the long term. Suitability for commercial production is the deciding factor in most current orchid growing.

The basic hybrid *Oncidium* Nona, developed from the cross-pollination of *Oncidium crispum* and *Oncidium varicosum* at the end of the 1930s provided much material for the brilliant yellow-gold hybrids bred between 1960 and 1990. For the grower in intermediate or moderate subtropical climates, these developments provide limitless material without their having to add multigeneric derivatives.

I may be accused of promoting a monotony of colour in such a selection but, given that ondiciums and oncidium derivatives are usually subordinate in general collections, they warrant a mention here. This also applies to hybrids generated from the equitant section. But equitant oncidiums do not affiliate with the orchids in the list of genera already recognised as compatible with pseudobulb oncidiums.

Vandas and Associates

Apparently hybrid vandas occupied few places in orchid collections until about half-way through the twentieth century, when there was progress using species which had been cultivated since the early years of orchid growing. These species usually came into the hands of growers in small quantities through nurseries or auction rooms. As demand increased, vandas of all species obtainable were stripped from habitats in hundreds of thousands and packed for export.

Most large plants were bought by those with room to grow them. Descriptions of some cultivated plants in larger collections indicated that it was considered more important to have them flower profusely than to use them as seed or capsule carriers and pollinating clones.

The first known hybrid was *Vanda* Miss Joaquim, carrying the name of its raiser in Singapore. It stemmed from the pollination of *Vanda hookeriana* by *Vanda teres*. Both were terete-leafed types from a very warm and sunny climate, growing in almost open conditions in lowland Malaysia, Burma and Thailand. Exhibited by Sir Trevor Lawrence, in 1897, at its first appearance in Britain and Europe, it was awarded a First Class Certificate by the Royal Horticultural Society. The plant was over two metres high and the raceme carried twelve flowers, mostly unopened. (The history of this and another early hybrid is fully related in Growing Orchids Book Three, *Vandas, Dendrobiums and Others*.)

The second known hybrid was *Vanda* Marguerite Maron, obtained from the cross-pollination of *Vanda suavis* and *Vanda teres*, flowered by C. Maron in France in 1903. This was the combination of two different types, one with terete foliage and the other, *Vanda suavis*, belonging to the 'strap-leaf' group. The first and second registered hybrids were the only two known to Rolfe and Hurst when they produced *The Orchid Stud Book* in 1909.

Vanda coerulea, as is recorded in Growing Orchids Book Three, *Vandas, Dendrobiums and Others*, was discovered by the botanist Hooker about 1850 and introduced shortly afterwards. But it was not used as a pollen or seed parent until 1919, when the French grower Jean Gratiot used it as the capsule-carrying parent with *Vanda tricolor* to produce the hybrid *Vanda* Gilbert Triboulet.

It was also used with *Vanda suavis* to produce a hybrid registered two years later and named *Vanda* Herziana, which proved to be a duplication of a known natural hybrid. This sort of thing was rather common in early hybridising. *Vanda* Moquettiana came from another cross-pollination by a European grower using the same parents, but it is listed as synonymous with *Vanda* Herziana.

In 1931 a peak in orchid breeding was reached when *Vanda coerulea* was used as a capsule-carrying parent with *Vanda* (*Euanthe*) *sanderiana* as the pollen parent. The resultant grex was named *Vanda* Rothschildiana by Chasaing, head gardener of one of the Rothschild establishments. He must have occupied a very senior position and made the cross-pollination or raised the seed and flowered the seedlings.

The Rothschild family contributed greatly to orchid cultivation and hybrid production, but their horticultural interests were fairly general. *Vanda* Rothschildiana has been responsible for the greater part of hybrid production in the genus and its associates in the twentieth century.

A Philippines orchid of great variety and beauty, *Vanda sanderiana* is still referred to as a vanda in hybridising, although it has been transferred to the genus euanthe. The flower varies from brownish red to pink, and it occurs as an albino. In hybridising these differences were used to attain multi-coloured cultivars (see Growing Orchids Book Three, *Vandas, Dendrobiums and Others* and elsewhere).

It would be remiss to ignore the part played by *Vanda coerulea*, because most hybrids emerged from its inclusion in breeding *Vanda* Rothschildiana. Progress was slow, with very few noteworthy grexes following quickly from this outstanding orchid. It should also be acknowledged that the best flowers came principally from later horticultural varieties of both species raised from the cross-pollination of attractive scions. An illustration of a 'new look' *Vanda sanderiana* appears in Growing Orchids Book Three, *Vandas, Dendrobiums and Others*. (The graph and information in Growing Orchids Book Two, *Cattleyas and Other Epiphytes*, page 55, indicates how this may be carried out, and the possible penalties which are attached to the system.)

The illustration on page 118 is a very early colour photograph of a grex imported into Australia in the 1930s, one of the first clones of cross-pollination of the two species.

Vanda sanderiana was used with *Vanda tricolor* to produce *Vanda* Tatzeri at the Prague Botanic Gardens in 1919, but *Vanda* Rothschildiana is given precedence because it had more use in generating later hybrids.

Although *Vanda coerulea* and *Vanda sanderiana* both carry patterns of reticulation on the flower parts, those of *Vanda sanderiana* are more pronounced on the lower sepals than other areas and this is noticeable in nearly all descendants of the species.

The middle years of the twentieth century were regarded as the golden period of cross-pollination, using derivatives from all known species, but with the prime material obtained from *Vanda sanderiana*. Hybridists were not confined to any particular country, but Asians and Americans originated most of the cross-pollinations.

The genus in its hybrid forms was regarded by some as a replacement for the ubiquitous cymbidium, but that genus took on a new look and hybridists developed new forms, the first of which was a miniature series.

Although definite colouring stamps *Vanda sanderiana* derivatives, the larger flowers developed from this source had intense gold and blue infusions brought prominently into sepals and petals. The blue came from *Vanda coerulea* and the golden tints from other sources.

The number of vanda cross-pollinations registered with the Royal Horticultural Society and recorded in *Sander's List of Orchid Hybrids* had grown from fewer than 100 in the 1945 edition to many times that number by the 1990s.

In the last forty years of cross-pollination there was recourse to the species used originally, particularly in the instance of *Vanda* Rothschildiana and its progenitors, *Vanda coerulea* and *Vanda sanderiana.* The motives of hybridists were easily identified because the qualifications for top ranking hybrids were a rounded outline, a flat profile and overlapping petals and sepals. The genus as a whole had no restrictions or demands on the shape or size of the labellum, in contrast to the requirements for 'cattleyas', other composite genera or cymbidiums.

Colour development was taken to extremes in blue or red, mostly with adequate allowance for darker reticulation through these colours. At no time were clarity of colour or other criteria expected of hybridists, and some flowers with awards are grossly unclear. Some of the characteristics sought are unattainable, such as longer racemes and better spacing of flowers. Most of these are governed by the original characteristics of the species and none had much scope to spread out for accommodating larger flowers — these are typical of the unreasonable demands which spur modern hybridists to greater efforts.

The move into diversification followed the alignment of various genera — including a large number of Indo-Asian species — into compatible groups with vandas. They include acampe, aeranthes, aerides, arachnis, ascocentrum, ascoglossum, doritis, luisia, neofinetia, phalaenopsis, renanthera, rhyncostylis, sarcochilus, trichoglottis and vandopsis.

In theory it would be possible to combine all these into one hybrid form — something which is unlikely to occur in practice. It would probably depend on the order of combination and compatibility factors would need analysis plus a lot of guesswork.

The stage has been reached where vanda, arachnis, ascocentrum, phalaenopsis and vandopsis have been bonded to produce *Sutingara* Donny Low. Vanda, ascocentrum, neofinetia, renanthera and rhyncostylis have also been bonded to produce *Knudsonara* Rumrill Trinket. This created two five-genus complexes and brings together eight genera in two combinations. The next step would be to try cross-pollinating these two, but the operative word is 'try'. Nature being what it is, and chromosome numbers being the 'joker' in the pack, the chances of success are extremely slim.

Be that as it may, vandas have been brought into very beautiful, simple combinations that have had wide acceptance by orchid growers. Ascocendas, the combination of vanda and ascocentrum, are the principal of these. They are most numerous, although their base material is sparse; when looking at the ascocentrum content some inbreeding is present and the vanda content is apparently one-sided, represented principally by infusions of *Vanda sanderiana*.

It is not surprising that natural hybrids were recognised in the vandas collected and sent to Britain and Europe; four were listed in the first edition of *Sander's List of Orchid Hybrids*. It is surprising, however, to find that three were derived on one side from *Vanda coerulea*. This orchid grows in Northern India, Burma and Thailand. Frequently thought of only as a blue orchid, it varies from blue to white and, in Thailand, pinky-blue.

The hybrids identified were *Vanda* × *amoena* (*V. coerulea* × *V. roxburghii*, syn. *tesselata*), *Vanda* × *charlesworthii* (*V. bensoniae* × *V. coerulea*), *Vanda* × *confusa* (*V. coerulescens* × *V. parviflora*) and *Vanda* × *moorei* (*V. coerulea* × *V. kimballiana*).

Only *Vanda* × *charlesworthii* was used in hybridising in the first half of the twentieth century and a derivative, *Vanda* Oiseau Bleu followed in later lists. The other natural hybrids had limited use before 1900.

Changes in Nomenclature of Vandaceous Orchids

Some generic names were changed when hybrid lists were being assembled and validated for publication in *Sander's List of Orchid Hybrids*. These changes had little impact on vandas, but several saccolabiums, for example, became ascocentrums and rhyncostylus. Other genera were affected, too. It is apparent that ascocentrums and rhyncostylus were cultivated throughout orchid growing history, but in all that time little use was made of the ascocentrum or saccolabium pollen or seed-bearing capabilities and other vandaceous genera were similarly neglected.

Vandopsis, closely allied to vandas, were known as fieldia, stauropsis, grammatophyllum and vanda until assembled in their own small genus by Pfitzer and Schlechter in the late nineteenth and early twentieth centuries. Once used in hybridising, however, their original generic names remain effective for Sander's list. They have been used extensively since the 1950s.

Ascocendas

The combination of ascocentrums with vandas did not occur until the 1950s. Cross-pollinations of these small vandaceous orchids were probably made a few years before that, but registration of *Ascocenda* Portia Doolittle by Dr Sideris in 1949 was the first. Its parentage was *Ascocentrum curvifolium* × *Vanda lamellata*.

Only two or three ascocentrums form the basis of the group and *Ascocentrum miniatum*, a Thai orchid, when crossed with *Vanda* Rothschildiana as the pollen parent, produced *Ascocenda* Meda Arnold, which has remained the benchmark for excellence throughout the remainder of the twentieth century. The grex primarily responsible for its success was *Vanda* Rothschildiana, the importance of which was emphasised earlier.

Four hybridists were responsible for the first registrations between 1948 and 1960, three from Hawaii, the other from Thailand. The Thai cross-pollination was between *Vanda* Gertrude Myamoto and *Ascocentrum curvifolium*. *Vanda* Gertrude Myamoto originated from *Vanda sanderiana*, *Vanda dearii* and *Vanda tricolor*.

The development of the composite ascocenda population occurred principally in Honolulu, where it began, but this new breed was quickly adopted by Asian hybridists and it developed into a considerable number of registrations. In the late twentieth century ascocendas were produced almost world-wide and represented in many other combinations.

Kirsch of Hawaii, who raised *Ascocenda* Ophelia, another of the first six hybrids, used different material in *Vanda* Bill Sutton, but this again was a *Vanda sanderiana* derivative from its cross-pollination with *Vanda manila*. The pollen parent of *Ascocenda* Ophelia was also *Ascocentrum curvifolium*.

While the derivations could be followed through endless pages and become tiresome, the single species which emerges as outstanding is *Vanda sanderiana*. This Philippines species is illustrated in great detail and colour in *Orchidiana Philippiniana* by Helen Valmayor, with beautiful colour photography by Dick Baldovino.

Although ascocentrums are smaller than *Vanda sanderiana*, they contribute a miniature section to the complex and increase the number of flowers on some ascocendas. In this way they compensate for occasional deficiencies of the principal vanda progenitor. Their influence also intensifies colour and in some instances eliminates the reticulation which is the mark of some vandas.

Ascocendas enlarged the vandaceous composite into increasing numbers from the 1950s onwards and the various affiliates are outlined in the vanda section.

In a period when every effort was made to increase the size of hybrids it was unusual to find the reverse applying. The ascocendas are part of a perhaps unintentional miniaturising stage in vandaceous breeding, but that result is not unexpected, given the other genera bred into the complex.

Vandopsis and Miscellaneous Genera

All vandopsis were used in various cross-pollinations from the 1950s onwards, adding several new names to the list of vandaceous orchids. Although *Vandopsis parishii* has been transferred into the genus hygrochilus, it remains a vandopsis for hybrid registrations.

Grexes from cross-pollinations with vandopsis are tropical, rather than suited for cooler climate collections. In natural environments the plants grow as freely as their progenitors. An instance is *Renanopsis* Lena Rowold, illustrated on page 121, generated from *Renanthera storiei*, reputedly robust, and cross-pollinated with *Vandopsis lissochiloides*. Both are Philippines species (illustrated in Growing Orchids Book Three, *Vandas, Dendrobiums and Others*).

Vandopsis were not used to any great degree in hybridising. This is understandable when we consider *Vandopsis parishii*, because its flowers are sparse. *Vandopsis gigantea* carries as many flowers as *Vanda sanderiana* and they are substantial, but there are still only few to a raceme.

The producers of smaller associates of the vanda tribe gave dilettante growers flowers which were different yet never destined to achieve the popularity of the principal genera such as pure-bred vandas, cattleyas or paphiopedilums. They create focal points of cultivation which some find attractive because they are frequently isolated in cooler climates, however prolifically grown and flowered in semitropical or tropical climates.

In the last twenty years growers have tended to collect as many hybrids or species as will fit into the facilities available and to use them as groups attached to main genera. The orchids fall into three categories — cool growing, those suited to intermediate environments and those (the majority) suited best to subtropical or tropical climates.

These diversities vary from hybrids such as sartylis, a derivative of cross-pollination between sarcochilus and rhyncostylus, one of which is illustrated on page 123, and others from the almost limitless combinations registered in *Sander's List of Orchid Hybrids*.

The limit has almost certainly been reached in the number of affiliations possible in genera associated with vandas, but cross-pollinations into vanda species levels or combinations beyond the sixth tier have not ended. For commercial purposes, however, there must be limits to what can be attempted. Keeping stock on benches devoted to hybridising and sale wastes space unless it is continually renewed.

Most intergeneric hybrids of vandaceous origin are open-root productions which do not take kindly to container culture. When they must be grown in heated areas they take up too much space compared with smaller epiphytes. Hybridising and growing seedlings, apart from proving what is in production, is clearly limited. The only exceptions are derivatives from smaller generic combinations which are not so hungry for space and warmth.

Epidendrums

This is one of the oldest botanical nominations for orchids, conferred by Linnaeus in 1753. The derivation is from their habit and hosts — growing as epiphytes on trees. Epidendrums are among the genera which have been the playthings of various taxonomists in the nineteenth and twentieth centuries and few would equal its record in the numbers assigned to it or taken away.

It should be remembered that the generic term epidendrum was applied indiscriminately in the earliest periods of orchid taxonomy. Botanists had little idea of the number of plants and the extraordinary distinctions between them which would come to their notice in the eighteenth and early nineteenth centuries. Orchid taxonomy in its infancy added to lists gradually. Interchanges by authorities were frequent, mostly after painstaking minutiae had been considered and the determinations of other bodies reviewed. This continues and one is left either despairing or applauding!

The principal shift in epidendrum nomenclature concerned the transfer of many species from that genus to encyclia. The generic name encyclia was created by Hooker in 1828 and in the late twentieth century transfer of all species correctly belonging to the genus was incomplete. This should not have surprised orchid growers, because they probably would not have applauded wholesale reprinting of their favourite literature to complete the process. It was not possible to adopt the simple stance and say that all epidendrums with pseudo-bulbs should be encyclias.

It is surprising to find such a scarcity of names registered for epidendrum hybrids in early records. They were among the earliest cultivated orchid plants and it would be even more surprising for anyone to accept that early cultivators did not try to self-pollinate or cross-pollinate the flowers. After all, they did it to every other exotic plant which came into their possession. If they did so with epidendrums they probably stumbled over the pollinating process or growing the seed.

In the middle of the nineteenth century this process was poorly understood and failure was certain if growers were careless or failed to take simple precautions. Even more surprising was the chance germination of spilled seed which so many growers witnessed — seed scattered on benches, on pots containing other plants and even on the interior walls of glasshouses. Many incidents such as these appear in the records of the nineteenth century.

Among the earliest orchids I grew were two reed-stemmed epidendrums — one a species, the other a hybrid. They were always grown outdoors. The species plant never set a chance, insect-pollinated capsule, but *Epidendrum* Obrienianum, the hybrid, seldom missed one or two pollinations on the ever-lengthening flower racemes. The long life of the flowering head suggests that more than one sort of insect pollinated the flowers. Plants were raised by scattering seed on the surface material of the pot containing the fertile plant. Of many thousands of seeds scattered, only a few germinated and grew. The flowers resembled those of the parent plant, which I still have after more than fifty years. Perhaps hundreds of propagations have been taken from it and capsules still occasionally appear on the flowering stems.

Neither Dominy, Seden nor any other cross-pollinators of their time used epidendrums in their craft, according to Lewis Castle's book. However, soon after Castle had his work published, Seden's cross-pollination produced *Epidendrum* Obrienianum, the seed for which was sown, grown, flowered and named at Veitch's nursery in Britain in 1888. Veitch described the flowers of this new orchid as midway between the scarlet *Epidendrum radicans* and purple *Epidendrum evectum*. It is scarlet, but with little of the purple *Epidendrum evectum*, a species originally imported from Colombia.

Epidendrum radicans is one of the widespread easy-growing orchids of Central and South America and bears the colloquial epithet Crucifix Orchid, a term which has become attached to other similar hybrid flowers in the genus, particularly *Epidendrum* Obrienianum. Rolfe described it as 'very handsome and floriferous'.

The hybrids listed in Rolfe and Hurst's *Orchid Stud Book* were raised between 1888 and 1907 and many, of course, were between plants which have now been transferred into different genera, some more than once. The total cross-pollinations flowered and named up to publication and revision of their book in 1907 still amounted to only twenty-six. This number had increased to only thirty-nine almost fifty years later.

In the first half of the twentieth century epidendrums were cross-pollinated with brassavolas, cattleyas, diacrium (caularthron), laelias and sophronitis. Possibly the two most important cross-pollinations were between an epidendrum and a cattleya and an epidendrum and sophronitis. Those two hybrids survived where others have been lost in the inevitable shuffling that occurs in collections.

The cattleya cross-pollination survived for at least one basic reason. It has proved sterile in spite of many attempts by hybridists to further the relationship with other genera or species. The cross-pollination was between the capsule-carrying hybrid *Cattleya* Claesiana and *Epidendrum* Obrienianum. The product was *Epicattleya* Nebo, brilliant purple, with the shape of epidendrums. The habit of growth is reed-stem, in the correct environment growing to about a metre and a half (about five feet) tall. The flowers are illustrated in Growing Orchids Book Two, *The Cattleyas and Other Epiphytes*.

Since halfway through the twentieth century hybrid lists have been well supplied with cattleya-epidendrum cross-pollinations and some have been developed into forty or more different generic entities and in combinations of up to five different orchids.

In this hybridising process all semblance of the Crucifix Orchid has been lost and the general configuration is the accepted cattleya morphology and flower, as illustrations of the complex show.

A sense of humour is occasionally necessary for orchid growing. The reference to the first hybrid epidendrum, *Epi.* Obrienianum, and its parentage in the 1981–1985 volume of *Sander's List of Orchid Hybrids* has the female or seed-carrying plant as *Epidendrum erectum* instead of *Epi. evectum*.

The other noteworthy surviving cross-pollination was between *Epidendrum radicans* and *Sophronitis grandiflora*, best regarded as a synonym for *S. coccinea*. It was named *Epiphronitis* Veitchii, raised from a cross-pollination by Seden and flowered in 1890.

Other epiphronitis were raised in the late twentieth century and some proved non-sterile, while derivatives from *Epicattleya* Nebo have not appeared in the hybrid lists.

An extraordinary epidendrum cross-pollination was registered with *Epidendrum* Obrienianum as the seed parent and *Sophronitis* (*Sophronitella*) *violacea* as the pollen donor. This cross-pollination was made by E. O. Orpet of France, although the registration is credited to Thayer. The resultant hybrid was named *Epiphronitis* Orpeti in early catalogues, but it has not been noted since. Orpet also made the cross-pollination for *Sophrolaelia* Orpetii, which is still grown now.

Sophronitella violacea has not appeared as a genus with another orchid to produce a hybrid since 1901. It is Brazilian, a single species in a genus and one which I have grown for fifteen or more years. All my attempts at cross-pollination either way with sophronitis or sophrolaelia pollen have never appeared even momentarily to be fertile or present any chance of success.

Sketches of flowers of four pollinations which had epidendrums as parents, including *Epiphronitis* Orpeti, were sent to Oakes Ames, in America, who at least gave a clue to identifying this cross-pollination in the periodical *American Gardening*. It would seem from his report that the watering process carried out by Orpet was to dip pots on which seed was scattered up to the rim of the pot in water. From this it can be deduced that the seedling *Epiphronitis* Orpeti could have been a 'floater'. This was not new, as quite a number of instances had already been encountered in other seed-raising endeavours. The other three hybrids were derived from *Epi.* Obrienianum × *Epi. elongatum* (two forms) and *Epi. cinnabarinum* × *Epi. radicans*.

In the sketches of the four flowers, plus the information sent to him, Oakes Ames noticed little difference in the four, apart from variations of colour and minor shape changes. From his conclusions it would appear that the *Sophronitella violacea* seedling could have developed from a 'floater'. Yet there is little doubt that it was put forward in good faith. It could have been a poorly carried out cross-pollination and there is no information on what was sown on the seeded pot.

In considering the opinion of this American, more knowledgeable than most, one cannot help reflecting on the credibility of *Epiphronitis* Orpeti.

The first hybrid from a combination employing *Sophronitella violacea* is possibly yet to come, with this species either as seed capsule carrier or as the pollinator. It seems impossible that no species in the catalogue is compatible.

As for *Epiphronitis* Orpetii (as it should be spelt), having allegedly been done once, it should be possible for the cross-pollination to be repeated. There is a barrier, though not an insurmountable one, to consider: the flowering of the epidendrum is principally out of kilter with that of *Sophronitella violacea*, which is cool growing and flowers in winter, and the flowers are not long lasting. They wilt in a matter of ten days or a fortnight, much sooner if pollination is attempted.

It is interesting to follow early efforts to use *Epiphronitis* Veitchii as either seed or pollen parent. Many took up the challenge and one grower tried it with *Laelia purpurata*. He had two capsules set for four months, by which time they discoloured and were removed from the plant. The records indicate that they contained nothing but 'chaff'.

The second attempt was to pollinate three epiphronitis flowers with *Cattleya mossiae* and another three with *Cattleya mendelii*. Thirty other flowers of the epiphronitis were self-pollinated and proved barren, as they had done in previous years when flowering. The grower was noted as a hybridist in Britain and his achievements to prove the Mendelian theory were numerous and widespread over orchid genera, particularly paphiopedilums, of which he had a large collection.

In the early twentieth century the relative dominance of certain genera over others was given considerable attention. At a conference on this aspect and other facets of horticulture the doubts concerning *Epiphronitis* Veitchii were laid to rest, although orchids were not a principal consideration at the function.

It was disclosed in a paper on orchid hybrids that reverse cross-pollinations of the two species concerned in obtaining the hybrid epiphronitis had all failed and in most instances of cross-pollinations between epidendrums and other genera the pollen-parent epidendrums were dominant. *Epiphronitis* Veitchii is illustrated in Growing Orchids Book Two, *The Cattleyas and Other Epiphytes*.

Other epidendrum cross-pollinations which appeared early in the twentieth century included *Epi.* × Cleon (*Epi. pseudepidendrum* × *Epi. radicans*), *Epi.* Dellense (*Epi. radicans* × *Epi.* xanthinum), *Epi.* Gattonense (*Epi.* Boundii × *Epi. xanthinum*) and *Epi.* Kewense (*Epi. evectum* × *Epi. xanthinum*), the reverse cross-pollination of which also flowered about the same time.

If the first half of the twentieth century was poor in the number of epidendrum registrations, the years following issue of the first edition of *Sander's List of Orchid Hybrids* were full of innovation and intergeneric surprises. The years from 1960 to 1985 were also notable for epidendrum species transferred to other genera. Although explained, it made for confusion because one generic name was used in the literature and another in the hybrid list.

As an instance: *Epidendrum cordigerum* in the hybrid list became *Encyclia cordigera* in the literature, with synonyms adding to the confusion in older literature which refers to the orchid as *Epidendrum atropurpureum* or by other names. As a pagan or non-believer in the virtue of correctness, I always liked 'Dragon's Mouth' or 'Dragon's Claws', the native names used for the flower where it grew naturally.

The extent to which taxonomic changes were made can be recognised by the need to refer almost fifty epidendrum species in the 1980–1985 hybrid list to other generic or species titles in earlier lists. Most of these changes were necessary distinctions which should have been recognised many years previously. While every grower is deeply indebted to the producers of the hybrid list, they must study it to be aware of and use correct taxonomic designations.

In tracing the pedigrees of epidendrums it may be necessary to consult earlier editions of *Sander's List of Orchid Hybrids*. If so, it is best to go to the section for encyclias in the 1981–1985 edition to note which ones were transferred. There are thirty-six and they could be referred to either as encyclias or epidendrums in literature, depending on the year in which it was printed or the lack of direction in index or glossary. Sometimes synonymy is referred to, sometimes not.

To explain this in an easy way: if *Epicattleya* Filiberti Gibbet is noted in the 1981–1985 edition the parentage is given as *Cattleya* Dubiosa × *Epidendrum cordigerum*. The following entry is *Epicattleya* Florida, with the parentage *Cattleya dowiana* × *Epidendrum atropurpureum*. The correct name for both epidendrums is *Encyclia cordigera*. Not all are complicated in this way, but it indicates the differences which arise in nomenclature and why they occur.

All thirty-six epidendrums included in the 1981–1985 list have been used to create various hybrids in the book, from simple epicattleya cross-pollinations to complex hybrids such as *Rothara* Koolau Starbright, which has epidendrum, brassavola, cattleya, laelia and sophronitis in the pedigree. The epidendrum used was *Epi. phoenicium*, which has been removed to the genus encyclia and is now known as *Encyclia phoenicia*. Used in the middle years of the twentieth century, it was originally discovered in Cuba and named by Lindley in 1838, yet it was never a popular species and was not widely cultivated like others in the genus.

Reed-stemmed epidendrums have not had as much exposure to hybridising as other forms. In the middle years of the twentieth century a certain amount of work was carried out by hybridists, but the full potential of this group appeared to have been reached when the size of individual flowers was increased to twice that of the original species, principally *Epi. radicans*. The plants were so easily grown in warm environments that they virtually became garden plants.

In this way this orchid became indigenous in warmer areas of Australia and no doubt elsewhere in suitable regions where it was cultivated. It may not do this section of the genus justice to say that the dominant colours were red and yellow and they could be seen as stereotyped flowers without the variety of more intergeneric intrusions. They created varieties of easily grown plants, as cymbidiums did in intermediate climates.

Orchid growers who are confused should not think they are alone! But if the *Handbook on Orchid Nomenclature and Registration* is available it will help to explain many problems which appear difficult. Epidendrum is not the only genus affected, but it is so widespread throughout North, Central and South America that it may take many years for transfers to be completed. As far as hybrids are concerned, the epidendrums, once used and so named in hybrid parentage, remain epidendrums whether they are known botanically as encyclias or by any other name.

The work of epidendrum and encyclia hybridising was divided among a considerable number of breeders in various countries and most occurred between 1960 and 1990. At least that was the period of greatest innovation and registration. The Hawaiian hybridist W. W. G. Moir was a prominent and avid worker, bringing into the list many innovations with intermixed epidendrum and encyclia cross-pollinations.

It appears that so-called 'cattleyas' attracted most attention and there was a drift towards miniature or smaller flowers rather than further development of the already well-placed laeliocattleyas and similar complex hybrids. The incidence of smaller hybrids generated by epidendrum and encyclia cross-pollinations was, of course, an unintended result caused by the initial flower size.

The breeding and production of miniature hybrids in which epidendrums and encyclias featured reached its apex early in the 1990s. The number and scope of combinations appeared to cover sufficient territory and this section of orchid hybridising virtually ceased.

Before leaving this interesting section, we should take a look at another instance of miscegenation. In one of the larger British collections in 1899 there was much jubilation because *Epidendrum* Obrienianum, cross-pollinated with *Dendrobium crystallinum,* had indicated pregnancy. The capsule grew, matured and shed seed, which was sown, grown and finally induced to flower. A contemporary publication recorded:

> They are practically reproductions of the mother plant; not quite identical with it perhaps and varying slightly between themselves, but without any recognisable trace of the pollen parent. This adds another to the series of seemingly inexplicable facts in connection with the hybridisation of orchids and we should like to see the experiment carefully repeated, or varied by using the pollen from some other dendrobium, the reverse cross being also tried. We may add that some of the plants were thrown away after the flowers appeared.

The answers are now quite plain. Either the pollination was inexpertly carried out or — something which could be explained had the epidendrum been a species — it was a parthenogenetic happening such as follows stimulation of other well-known orchids.

Laelias

Although the genus was considered in conjunction with cattleyas, it was subject to sufficient cross-pollinations within itself to demand attention, particularly in view of the cultivation of so many species brought into the glare of publicity by the conversion of orchid collections from hybrids to species — or elements of miscellaneous character, as they were so frequently described.

Hybridism in the genus occurred quite early in historic orchid growing and it is difficult to pinpoint early attempts or successful pollinations which occurred before registration in various British or European horticultural publications.

The best source for most of these is archival literature not readily within reach of most orchid growers. Fortunately for them, in the last years of the nineteenth century there were those who not only registered their own work but catalogued that of others who were less careful. Amongst the publications recording such cross-pollinations was that of Rolfe and Hurst, *The Orchid Stud Book*. It survived and became available again in the 1990s. Its principal value, apart from recording the names and dates of registrations, is that it lists the various publications in which the notices of flowering and breeding were included.

Until an enlarged catalogue of Brazilian laelias became known, between 1960 and 1990, appreciation of the genus was usually confined to those used in cross-pollinations with cattleyas. In those hybrids the use of smaller Brazilian species was limited.

The first hybrid using only laelias was said to have been raised in 1853 or 1854, one of Dominy's cross-pollinations. It flowered in the 1860s and two or three dates appear in records. The combination was between *Laelia crispa* and *Laelia perrinii* and the hybrid was subsequently named *Laelia* Pilcheri. It was recorded in the *Gardener's Chronicle* in 1865.

Some years elapsed before a follow-up in the genus produced *Laelia* Flammea from the cross-pollination of *Laelia cinnabarina* and *Laelia* Pilcheri in 1876. This secondary hybrid had brilliant flowers with somewhat 'spidery' outlines, but no more so than the majority of laelia pollinations produced.

Hybridising at the time appeared to concentrate on intergeneric cross-pollinations and it was the age of experiment. There was not much material to work on in laelias and it took a long time to obtain viable seed and successfully grow and flower the seedlings. Considering the rather rough techniques compared with those of today, the methods of 100 years earlier rendered much of the seed scatter useless. Only a small proportion of flowering plants were produced from the available seed.

Apparently an effort was made to freely use *Laelia pumila* and *Laelia tenebrosa*, and of thirty-three cross-pollinations in the late nineteenth and early twentieth centuries *Laelia pumila* appeared in five and *Laelia tenebrosa* in eight. Only sixteen other laelias appeared, including hybrids, an indication of how many species were in circulation at the time.

During the fifty years following publication of the Rolfe and Hurst list in 1900, only fifty-three additions were made. Many of these again used the principal species of the first list and surprisingly few secondary hybrids were generated from that list.

Natural hybrids in the overall population of the genus were rarely noted and those which came under scrutiny in early history were principally from the Central American region. Four came from that region: *Laelia* Finckeniana (*L. albida* × *L. grandiflora* syn. *L. speciosa*), *L.* Eyermanniana (*L. albida* × *L. anceps*), *L.* Leucoptera (*L. albida* × *L. furfuracea*) and *L.* Venusta (*L. furfuracea* × *L. grandiflora*). Two came from Brazilian species: *L.* Wyattiana (*L. boothiana* × *L. crispa*), and *L.* Caetensis (*L. flava* × *L. crispata*).

This by no means exhausted the possibilities of finding other natural cross-pollinations, which could be expected from among the large number of Brazilian laelias which have come into cultivation since the 1950s. Many of these were wrongly identified when exported to other countries. The blame for this should not be laid at the feet of botanists and naturalists in Brazil. It was caused instead by poor appreciation and dissemination of information by government instrumentalities and lack of adequate communication between horticultural bodies.

The richness of this section of Brazilian flora was probably unsuspected until such publications as *Orchidaceae Brasilienses*, from Pabst and Dungs, brought its possibilities into prominence. The Brazilian orchid flora was poorly represented in translation in many instances and was unfortunately subjected to severe reduction and destruction from about 1920.

As the systematic use of major material such as *Laelia purpurata*, *Laelia pumila*, *Laelia tenebrosa* and other Brazilian species is easy to follow in lists of cross-pollinations associated with cattleyas, use of these same species within their own genus must be sorted out from the mass of intergeneric hybrid registrations. When this is done it is clear that the bulk of the genus has been underused throughout its history.

This seems to show that either the Mexican *Laelia anceps* has been underused and undervalued or that there must be unacceptable features which affect cross-pollinations. This was remedied in some degree in the 1960s, when it was used as a parent twenty-nine times, a number exceeded only by *Laelia flava*, which entered the list thirty-seven times. *Laelia flava* is golden coloured and this may have been an influence in a period when gold was sought in miniature cross-pollinations.

If trends in orchid breeding are followed from Rolfe and Hurst's *Orchid Stud Book* through to *Sander's List of Orchid Hybrids* some patterns emerge for laelias and other small orchids which had no ready acceptance in the vogue for larger flowers such as laeliocattleyas and brassolaeliocattleyas. It was not until laelias indicated their great influence on colour breeding within their genus that they were accepted.

In hybridising in the 1960s they appeared in more than 200 cross-pollinations, within and outside the genus, using only eight species, three of which were earlier listed as major contributors.

The search for size as a desirable characteristic has been modified in the late twentieth century and although it still applies in some intergeneric hybrids, smaller laelia cross-pollinations within the genus were completely acceptable if the colour was good and the shape passable.

Other Brazilian natural hybrids were identified after those indicated in early lists. They included: *Laelia* × Carasana (*L. mantiqueirae* × *L. lucasiana*), *L.* × Cipoensis (*L. ghillanyi* × *L. crispata*), *L.* × Gerhardt-Santosi (*L. harpophylla* × *L. kautskyana*), *L.* × Espirito-Santensis (*L. pumila* × *L. xanthina*), *L.* × Leeana (*L. pumila* × unknown), *L.* × Lilacina (*L. crispa* × *L. perrinii*) and *L.* × Wyattiana (*L. lobata* × *L. crispa*).

Many of these natural hybrids have not appeared in cross-pollinations, but it should be recognised that not all cross-pollinations go to the authorities responsible for keeping the orchid community informed and publishing the names in editions of *Sander's List of Orchid Hybrids.* Identification sometimes depends upon verification by raising and flowering manual cross-pollinations and comparing them with originals or original descriptions. This is somewhat difficult with Brazilian laelias in particular, because there appear to be intermediate forms and varieties of several well-known species.

The additional names were taken from *Orchidaceae Brasilienses* and this is perhaps the best of later authoritative publications to consult. Other authors may disagree with the nomenclature, but revisions are very hard to follow in recent literature. Lack of communication between various countries may also make criteria difficult to accept. Much more importantly, I have never met growers or others more qualified who are able to correctly interpret descriptions of flower parts without illustrations or botanical specimens to guide them.

Assessing the Brazilian laelias from such authoritative works as *Orchidaceae Brasilienses* frequently involves fine points such as the shape or length of the labellum and inequalities of parts such as petals or sepals. Colour is indeterminate where some species may vary from white through pale yellow and pink to rose and deeper colours. Morphology may also be confused, with different forms of pseudo-bulbs, leaf length and flower stem length.

In all, it appears to untrained eyes that the Brazilian laelia family is large and diverse and not necessarily correct in its nomenclature. Synonymy or varietal status appears a valid proposition for some species. Few hybrids have been developed from the large list and it seems that moves toward cross-pollinations are frequently spurred by challenge rather than having definite objectives.

Considering the extent of hybrid lists published each month in the *Orchid Review*, particularly in relation to their diversity, it is apparent that more dilettante workers are entering the hybridising field each year and the number of each grex entering collections is therefore small.

Brazil was fortunate in the number of orchid botanists who worked there, their expertise and the literature they produced. The field was well endowed with material, but unfortunately much of it is lost to the English-speaking communities because it has not been translated. This applies particularly to such authors as F. C. Hoehne, whose beautifully illustrated books are almost wasted outside Spanish-speaking areas.

Cymbidiellas

Cymbidiellas are not unique among genera on which no work was carried out in cross-pollinating until the late twentieth century. There are hundreds of perhaps minor and unimportant members of the orchid world on which hybridising has not been tried. In this instance, however, the neglect is surprising. There was a considerable time lapse between the discovery and cultivation of cymbidiella species and the generation of hybrids, yet the species are very attractive, striking orchids and it is possible many ineffective and sterile cross-pollinations have been tried.

The species first brought into cultivation in Britain was probably *Cymbidiella humblotii*, which was flowered in 1892. But the genus was known and catalogued about seventy years earlier in Madagascar by Thouars when he found *Cymbidiella flabellata*. He described the flowers as green, purple and yellow. The outline drawing in his book *Histoire particulière des plantes orchidées recueilliés sur les trois îles australes d'Afrique* is indicative of the cymbidium-like flowers carried on a slender stem with the blooms opening consecutively upward. The raceme or spike also indicates in photographs of a later period that it elongates as the lower flowers wither.

Cymbidiella flabellata was included in *Sander's List of Orchid Hybrids* in a 1982 registration by Dr Martin Orenstein, an American, when it was the pollen parent used with *Cymbidiella rhodochila*. The registrant named the hybrid *Cymbidiella* Kori Dingeman.

The genus is also compatible with eulophiella, another Madagascan orchid. Although morphologically similar to cymbidiums, it is apparently not compatible with them, despite earlier listing as a cymbidium. Rolfe created the genus cymbidiella in 1918. The name indicated a diminutive of cymbidium.

The correct botanical name for *Cymbidiella rhodochila* is *Cymbidiella pardalina*. There are only three known species and each has a completely different habitat and mode of growth. The flowers of *Cymbidiella rhodochila* are striking and brilliantly coloured, as the illustration on page 127 shows.

When infrequently exhibited the species evoked much admiration and in 1905 the Royal Horticultural Society orchid committee awarded a First Class Certificate to a plant of *Cymbidiella rhodochila* exhibited at a meeting of that society. One can well imagine the pride with which it was presented to the orchid committee members.

A considerable amount of exploratory work awaits someone willing to further hybridise these rather obscure species. That also means analysing the environments in which they grow and the obvious necessity for controls so that each species is suited. It will probably be found that primary hybrids could be better to work with once the initial problems of setting capsules and raising seedlings are overcome.

Cultural recommendations for cymbidiellas are hard to define, but the kinds of environments designed for such orchids as phalaenopsis would be suitable. Because it is unlikely that many plants would be available unless seedlings are bought and successfully grown, they must fit in with other orchids. It would be unwise to expect them to grow in climates with minimum temperatures throughout the year of less than 18 to 20 degrees Celsius (about 65 to 68 degrees Fahrenheit).

Eulophiellas

It was only in the 1970s that a successful attempt was made to breed into this genus from Madagascar. *Eulophiella rompleriana,* a species with morphological similarities to cymbidiellas, was the pollen parent to the seed-bearer *Cymbidiella rhodochila.* The capsule proved fertile and the seedlings were probably raised and named by the registrant, Dr Orenstein, in 1981. The genus eulocymbidiella (with the contracted title eucmla) was created for the hybrid, which was given the name *Eucmla* Susan Orenstein. This intergeneric hybrid preceded by a year the intrageneric cross-pollination named *Cymbidiella* Kori Dingeman.

The growth habit of both genera is rambling and not easily accommodated in containers other than slat baskets. For *Eulophiella rompleriana* it would need to be quite large, at least half a metre square, not necessarily deep, but allowing for a rambling, coarse root system somewhat like that of vandas or phalaenopsis.

This system indicates what is known as open-root cultivation, with plenty of scope to travel in coarse material used for filling and perhaps additional allowance for pendant roots to develop. Growers often use containers when more open systems such as slabs of fern, bark or wood mounts or slat baskets of wood are essential for cultivating certain genera.

The need for a moist, warm atmosphere and fairly bright light is indicated by these restrictions, which may not necessarily suit phalaenopsis. Cymbidiellas or eulophiellas may also suffer burning unless shading is correct but not too dense.

Further development and introduction of other species of both genera is possible from the beginnings so long delayed after both were first known and cultivated, but hybrids are likely to remain scarce.

Interest in eulophiella was shown by Japanese hybridists and a cross-pollination using the genus appeared in the hybrid lists for 1988. *Cymphiella* Hiroshima Peace (*Cymbidium pumila* × *Eulophiella rolfei*) created another addition to the steadily increasing number of intergeneric hybrids. Depending on their attractiveness, they may be elaborated into worthwhile 'breaks' in cymbidium-like orchids.

Conclusion

This book gives only a glimpse of the enormous list of hybrid orchids registered in publications of the twentieth century, but those dealt with reflect the major sections and briefly describe the trends from the early essays in cross-pollination to the multiplicity of generic combinations of the 'tertiary phase' of the cult.

The effect of meristeming or clonal reproduction had an effect on orthodox propagating and seed-raising which is almost impossible to assess, but it made available reproductions of most hybrids in a faster time than was possible by ordinary propagation. Some genera, such as paphiopedilums, showed resistance to the technique and in the 1990s still defy mericloning. This fact does not, however, prevent demand for clones derived by ordinary propagating, if they are ever available.

This stubbornness in resisting clonal reproduction affected paphiopedilums in a way not followed by other genera and caused a return by growers to species level. The generation of older forms of hybrids, with reworked colour and derogation of the rounded shape sought and obtained after so many years of work, was a natural adjunct of the phase.

The hybrid forms of the genus paphiopedilum were unique in this. The new was exchanged for the old and created a trend in growing which spread to areas never before thought possible. There was a demand for hybrids which would withstand subtropical to tropical climates. This was in direct contrast with ideas held in the mid-twentieth century about growing hybrid paphiopedilums produced from Indo-Asian stock.

This Indo-Asian type hybrid was developed to perfection according to preconceived standards developed over the years from the 1930s onwards. Yet these standards were probably envisioned in the earliest years of cross-pollinating. The hybrid lists published after the 1950s were overflowing with new registrations, most of which followed the route to a rounded outline, each one different even when raised from the same capsule.

But a new breed emerged, developed from species originating in areas such as equatorial Pacific and Asian regions, with a tolerance for warmth not shown by Indo-Asian types. Fashion dictated that growers of paphiopedilums follow leads in new directions dictated by use of several species discovered or rediscovered late in the twentieth century. Styles of flowers had changed and with those changes the appreciation or award systems were also altered.

The future of the genus is difficult to predict, but no doubt it will follow the pattern set by other genera. The orchid-growing population increased greatly in the last half of the twentieth century and although most of the increase followed paths that were well trodden before, they have infinitely more variety in paphiopedilums to tempt them.

There are no intergeneric distractions to compromise the integrity of paphiopedilums and for someone who, like the author, has seen hybrid development for much of the twentieth century, it appears that all the permutations have been made and little is left but variation.

Although phalaenopsis have been interbred with other genera, in essence they resemble paphiopedilums in their insularity and in the numbers cultivated. The genus in hybrid form came from few species and it was only in later years of their breeding that deviation occurred into smaller and perhaps more brilliant forms to enhance the variety and add imagined or real characteristics to orthodox flower colours and forms, which were principally pink to white and moth-like.

It may appear to casual observers that little has been added to phalaenopsis hybrids in the last fifty years other than a vast increase in the number of registrations. In point of fact, the number was not matched by a variation of basic colour and shape. There is a preponderance of white in the listed hybrids, with minor alterations to labellum shape and colour. It is difficult to deny, however, an entitlement to add a new name to the list for a new flower which completely or almost touches 'perfection', as outlined in whatever standards are applicable, even if it is one among thousands which look alike.

For example, in *Sander's List of Orchid Hybrids* for 1981–1985, the entry for hybrids generated from *Phalaenopsis amboinensis* added 155 new names to the list. Although there may have been considerable differences between the pollen parents used to generate seed capsules on *Phalaenopsis amboinensis* it is possible that many of the hybrids flowering were insufficiently distinguishable from each other.

Phalaenopsis comprise a major section of *Sander's List of Orchid Hybrids* and to casual observers looking at the genus collectively in flower there is little difference between many of the hybrids. The principal difference is in the final production of the flowers and their compliance with certain standards of excellence. The proportion which comply with these standards is minuscule, but as with 'cattleyas', what grower could put an orchid plant in the garbage?

Gradations in colour, verging from pink through to deep rose-purple, generated by the addition of doritis or coloured forms of species such as *Phalaenopsis lueddemanniana*, have been common for about fifty years and the aim of hybridists to bring in other colours has been partly rewarded. In the 168 pages of registrations for phalaenopsis for 1981–1985, however, the principle of registering all submissions must be questioned. Those 168 pages do not include all the intergeneric cross-pollinations. This problem must be faced squarely in relation not only to phalaenopsis and its affiliates, but also for other genera. This may appear rather difficult to appreciate, particularly to those who enter orchid breeding in the twenty-first century. In this regard it should be noted that the list of registrations has reached a stage impossible to continue in its present form. These are some of the problems facing registering authorities, centred at present on the Royal Horticultural Society in Britain. Registrants must realise that if fees rise the charges are more than justified. It could become necessary to make them almost prohibitive and cause growers to think twice.

Most worthwhile hybrid orchids have been processed through ordinary propagation and finally through clonal propagation. Some cloning has been less than satisfactory from the point of view of proliferation beyond first-generation mericlones, mostly because of poor methods and material. An example of this appears on page 109 of *The Specialist Orchid Grower* in the illustrations of mericlone flowers of *Brassolaeliocattleya* Malworth 'Orchidglade'. These were three of eighteen photographs, no two of which were alike.

Experience of this kind of disappointing result has moved some growers to prefer seedling stock when renewing their collections rather than selecting clones of hybrids from catalogues, however good the photographs portraying the mericloned merchandise. The few bad results condemn the good ones, which are the majority.

Even the innovation of chemical treatment of cloned extractions proved reliable in only a few instances where conversion of ploidy was achieved. There was a lot of waste material.

In looking towards the twenty-first century imagination can run riot with the predictions of enthusiasts. The fashionable trend is to look at genetic engineering for the introduction of colours unknown in either original or hybrid flowers. If it was possible to introduce a gene for blue in cymbidium mericlones or seedlings the originator could look forward confidently to a rosy future rather than a blue one! In point of fact, considering the current glut on the market of plants of this genus, it is the only phase not reached since hybridising began and is a natural conclusion to a span of about 100 years of cross-pollinating.

In the middle of this period, some very notable clones giving promise of breakthroughs in colour were discarded for the simple reason that the 'will o' the wisp' of shape blinded hybridists to the possibilities inherent in the product in their hands. Many prospective cymbidiums were bred from stock long since gone beyond recovery. The rich purple of *Cymbidium* Cordelia 'Kuringai' comes to mind, an Australian clone of execrable shape but magnificent colour. There were others, one of which, *Cymbidium* Joyance 'Cinnamon', is illustrated on page 78. Little use was made of either of these two, even if we allow for the fact that their ploidy may have been a barrier.

In orchid breeding the paramount desire is possibly for blue flowers in the genera which do not have that content in their genetic make-up. Modern techniques make this introduction possible but, so far, to breed a blue 'cattleya' or a blue odontoglossum has proved impossible. Blue tints have been achieved, but no intensity of colour.

It is now up to the geneticists to have their try. There is an initial market for commercial production and when that is attained and every grower who wants a blue 'cattleya' or cymbidium has a clone in his or her collection we will have really reached the end of the road.

Bibliography

The following books and papers from the private library of the author were used as reference material in writing the seven books of this series:

Periodicals

Australian Orchid Review, Printcraft Press Pty Ltd, Sydney, 1936–1990.
Orchadian (journal of the Australasian Native Orchid Society), A.N.O.S., Sydney, 1983–1988.
Orchid Review (Britain), Royal Horticultural Society, England, 1893-1990.
Orchid World, vols. 1–6, 1910–1916 (publication discontinued).

Books

Anderson, Frank J., *Illustrated Treasury of Orchids* (reproductions of old lithographs), 1979.
Arcadio Arosemena etc., *Orchideas de la Costa del Ecuador*, Associacion Ecuatoriana de Orchideologia, Guayaquil, Ecuador, 1988.
Asia Agri-Business Corp., *Phalaenopsis Kingdom*, Agri-Business Corp., Formosa, 1987.
Associacao Orchidofila de Sao Paulo, *Native Orchids of Brazil*, 1977.
Australasian Native Orchid Society, *Checklist of Australian Native Orchid Hybrids*, 1981.
Australian Orchid Foundation, *Australian Orchid Research*, 1989.
Australian Orchid Foundation, *The Orchidaceae of German New Guinea* (translated from German), 1982.
Ball, John S., *Southern African Epiphytic Orchids*, Conservation Press, Johannesburg and London, 1978.
Bateman, James, *The Orchidaceae of Mexico and Guatemala* (originally published 1837), Johnston Reprint, Amsterdam.
Bedford, R., *A Guide to Australian Native Orchids*, Angus and Robertson, Sydney, 1969.
Birk, Lance A., *The Paphiopedilum Growers' Manual*, Pisang Press, Santa Barbara, Cal., U.S.A., 1983.
Braem, Guido J., *The Brazilian Bifoliate Cattleyas*, Brücke-Verlag, Kurt Schmersow, Hildesheim, Germany.

Castle, Lewis, 'Orchids, Their Structure, History and Culture', *Journal of Horticulture*, London, 1887.

Chow Cheng, *Formosan Orchids*, Chow Cheng Orchids, Taichung, Taiwan, 1970.

Clements, M., *Preliminary Checklist of Australian Orchidaceae*, National Botanic Gardens, Canberra, 1982.

Colombia: A Reference Guide, issued by the 7th World Orchid Conference, 1974.

Cooper, Dorothy, *New Zealand Native Orchids*, Price Milburn, Wellington, N.Z., 1981.

de Oca, Raphael Montes, *Humming Birds and Orchids* (illustrated by de Oca; translated by N. P. Wright), Editorial Fournier S.A., Mexico,1963.

Dockrill, A., *Australian Indigenous Orchids*, The Society for Growing Australian Plants, Sydney, 1969.

Dodson, C. H., and Gentry, A. H., *Flora of Rio Palenque*, The Marie Selby Botanic Gardens, Florida, U.S.A., 1978.

Dressler and Pollard, *The Genus Encyclia in Mexico*, Associacion Mexicana de Orchideologia, Mexico, 1974.

Du Petit-Thouars, Aubert-Aubert, *Histoire particulaire des plantes orchidées* 1882, Reprint E. M. Coleman, New York, 1980.

Dunsterville, G. C. K. and Garay, L. A., *Venezuelan Orchids*, Deutsch, London, 1959–76 (6 volumes).

Encyclopaedia Brittanica, 1980.

Fawcett, W. and Rendle, A. G., *Orchids of Jamaica*, Bishen Sing Mahendra Pal Singh, Dehra, India, 1982 reprint.

Fitzgerald, R. D., *Australian Orchids*, 2 vols, Lansdowne Editions, Sydney, 1977 reprint.

Fowlie, J. A., *The Brazilian Bifoliate Cattleyas*, Day Printing, California, U.S.A., 1977.

The Genus Sophronitis, 1987 (Orchid Digest Reprint).

The Genus Lycaste, Day Printing, California, U.S.A., 1970.

Gale, J. A., *Orchids of Cuba*, Bishen Singh Mahendra Pal Singh, Dehra, India (1987 reprint).

Garay, Leslie A., *The Genus Phragmipedium*, 1979 (Orchid Digest Reprint).

Halbinger, F., *Odontoglossum in Mexico and Central America* (Orchidea), Associacion Mexicana de Orchideologia A.C., 1987.

Halle, Nicholas, *Flor de la Nouvelle Caledonie et Dependences, 8: Orchidées*, Musée Nationale d'Histoire Naturelle, Paris, 1977.

Hamer, Fritz, *The Orchids of El Salvador*, The Marie Selby Botanical Gardens, Florida, U.S.A., 1981.

Handbook on Nomenclature and Registration, International Orchid Commission, London, 1985.

Hawkes, Alex, *Encyclopaedia of Cultivated Orchids*, Faber and Faber, London, 1965.

Hoehne, F. C., *Flora Brasilica, Vol. 12, Orchidaceas, Parts 1 to 4*. Companhia Brasiliera de Impressao e Propaganda, Sao Paulo, Brazil.

Holttum, E., *Flora of Malaya*, 1987 (Reprint).

Japan Orchid Growers' Association, *Quality Stream 'Cattleyas' — The 800 Collection*, Japan Orchid Growers' Association, 1985.

Jayaweera, D. M. A., *Flora of Ceylon*, 1893–1900.

Johns, John and Molloy, Brian, *Native Orchids of New Zealand*, A. H. and A. W. Reed, Wellington, New Zealand, 1985.

Johnson's Botanical Dictionary, 1917.

Kupper, Walter, and Linsenmaier, Walter, *Orchids*, translated by Jean W. Little, Nelson, London, 1961.

Laseron, Charles, *The Face of Australia* (first and revised editions), 1927–1949.

Lavarack, P. D., and Gray, B., *Tropical Orchids of Australia*, Nelson, Melbourne, 1985.

Lecoufle, Marcel and Rose, Henri, *Orchids*, La Maison Rustique, Paris, 1957.

Lindley, John, *Sertum Orchidaceum*, 1938 (Reprint).

Mark, Fred, *Introduction to Chinese Slipper Orchids*, 1987 (Orchid Digest Reprint).

Mark, Fred, Ho Fu Shun, Fowlie, J. A., *Cymbidiums of Taiwan*, 1986 (Orchid Digest Reprint).

Menezes, L. C., *Cattleya Labiata (Lindley)*, 1987 (Orchid Digest Reprint).

Missouri Botanic Gardens, *Orchids of Panama*, 1980.

Morris, Brian, *Epiphytes of Malawi*, Mardon Printers, Rhodesia, 1970.

Nicholls, W. H., *Orchids of Australia*, Nelson, Melbourne, 1969.

Oakes Ames and Correll, D. S., *Orchids of Guatemala*, Bishen Singh, Mahendra Pal Singh, Dehra Dun, India, 1983.

Ortiz, Pedro, and others, *Orchideas Ornamentales de Colombia*, Carlos Valencia Editores, 1980.

Pabst, G., *Rupicolous Laelias*, 1984 (Orchid Digest Reprint).

Pabst, G., and Dungs, F., *Orchidaceae Brasilienses*, 2 vols, Brücke-Verlag Schmersow, Hildesheim, Germany, 1975-77.

Paxton's Botanical Dictionary, 1868.

Pears' Cyclopaedia, 1900, 1930.

Perrier de la Bathie, H. *Flora of Madagascar*, Stephen D. Beckman, (translation, reprint) 1981.

Pocock, Maynard R., *Ground Orchids of Australia*, Jacaranda, Milton, Queensland, 1972.

Pradhan, Udai C., *Indian Orchids, Guides 1 and 2*, Udai C. Pradhan, Kalimpong, India, 1976, 1978.

Quisumbing, Dr Eduardo, *Philippine Orchids*, Eugenio Lopez Foundation Inc., Manila, 1981.

Rentoul, J. N., Growing Orchids Book One, *Cymbidiums and Slippers*, Lothian, Melbourne, 1980.

Growing Orchids Book Two, *The Cattleyas and Other Epiphytes*, Lothian, Melbourne, 1982.

Growing Orchids Book Three, *Vandas, Dendrobiums and Others*, Lothian, Melbourne, 1982.

Growing Orchids Book Four, *The Australasian Families*, Lothian, Melbourne, 1985.

Growing Orchids, *The Specialist Orchid Grower*, Lothian, Melbourne, 1987.

Growing Orchids, *Expanding Your Orchid Collection*, Lothian, Melbourne, 1989.

Rolfe, R. A. and Hurst, C. C. H., *The Orchid Stud Book*, Frank Leslie and Co., London, 1909.

Sagarik, Rapee, *Beautiful Thai Orchids*, the Orchid Society of Thailand, Bangkok, 1975.

Sander, Frederick, *Reichenbachia*, 1888, Reprint, 4 volumes.

Sander's List of Orchid Hybrids, Royal Horticultural Society, London, 1945–85, 6 volumes.

Sander's Orchid Guide, St Albans, UK, 1927.

Sauleda, R, and Adams, R., *Bahama Package*, reprint of papers, 1982-1987.

Saunders, W. Wilson (ed.), *Refugium Botanicum* (1869–1872–1882), Reprint 1980.

Schlechter, Rudolf, *African Angraecoid Orchids*, Australian Orchid Foundation, 1986.

The Orchids of Micronesia, (1914, 1921), Australian Orchid Foundation, 1986.

Flora of New Caledonia, Australian Orchid Foundation, 1986.

Orchid Flora of Sumatra (1911), Australian Orchid Foundation, 1986.

Schelpe, E., *An Introduction to South African Orchids*, Macdonald, London, 1966.

Schultes, R. E., and Pease, A. S., *Generic Names of Orchids*, Academic Press, New York, 1963.

Schweinfurth, Charles, *Peruvian Orchids*, Chicago Natural History Museum, 1958–70, 4 volumes with supplement.

Seidenfaden, Gunnar, *Orchid Genera in Thailand: Dendrobium Sw.*, Opera Botanica, Copenhagen, 1985.

Smith, J. J., *Die Orchideen von Java*, 1984, Reprint.

New Papuasian Orchids (translation), Australian Orchid Foundation, 1986.

Som Deva and Nathani, H. B., *The Orchid Flora of North-West Himalaya*, Print and Media Associates, New Delhi, India, 1986.

Sprunger, S. (ed.), *Curtis's Botanical Magazine Reproductions*, Cambridge University Press, London, 1986.

Stewart, J. and Hennessy, E. F., *Orchids of Africa*, Macmillan, London, 1981.

Stewart, Joyce, and others, *Wild Orchids of Southern Africa*, Macmillan, Johannesburg, 1982.

Valmayor, Helen, *Orchidiana Philippiniana*, Eugenio Lopez Foundation Inc., Manila, 1985, 2 volumes.

van der Pijl, L. and Dodson, C. H., *Orchid Flowers, Their Pollination and Evolution*, University of Miami Press, Florida, 1966.

Veitch, James, *A Manual of Cultivated Orchidaceous Plants*, H. M. Pollett and Co., London, 1887–1889, 2 volumes (original editions).

von Drateln, Herman, *Odontoglossum Culture*, Himalaya, Mexico, 1960 (?).

Warner, R. and Williams, B. S., *The Orchid Album*, B. S. Williams, London, 1883-1885, vols 2, 3 and 4.

Waters, V. H. and Waters, C. C., *Survey of the Slipper Orchids*, Carolina Press, North Carolina, U.S.A., 1973.

Wiard, Leon A., *An Introduction to the Orchids of Mexico*, Comstock Publishing Associates, Ithaca U.S.A., and London, 1987.

Williams, Brian and others, *Orchids for Everyone*, Herbert Michelman Books, New York, 1980.

I also referred to proceedings of various orchid conferences for useful information and data, as well as various publications of Alfred Russell Wallace.

Reference maps used were *South America*, by Kummerley and Frey; *The Philippines, Malaysia, Vietnam and Other Asian Countries*, by Robinson; *Bangladesh, Sri Lanka, including Burma*, by Bartholomew. Outdated references to boundaries and state changes were traced through the older encyclopaedias.

In consulting these and other references I noted many conflicting facts and found many mistakes, including some in what are usually considered impeccable sources. These were accepted as part of the processes of recording, interpretation and, not least, the printing and reproduction, and not held too unkindly against the authors.

References from the *Orchid Review*, edited by Robert Allen Rolfe and others, into the *Gardeners' Chronicle* were made available by courtesy of the National Herbarium, Melbourne.

The illustrations in this book were selected from thousands of transparencies in the author's library. The photography was carried out by the author, except where acknowledged to be by others. The cameras used over about forty years were principally three Pentax models, up to the 1989 series SFx, and others included a Russian copy of a Leica, which was the first; a Russian Fed (52 mm), a Praktisix, and a Marniya C45. The films used were principally Kodak, Ektachrome or Agfa.

Index